INVESTIGATIONS IN NUMBER, DATA,

Multiplication and Division

Things That Come in Groups

Grade 3

Also appropriate for Grade 4

Cornelia Tierney
Mary Berle-Carman
Joan Akers

Developed at TERC, Cambridge, Massachusetts

Dale Seymour Publications

The *Investigations* curriculum was developed at TERC (formerly Technical Education Research Centers) in collaboration with Kent State University and the State University of New York at Buffalo. The work was supported in part by National Science Foundation Grant No. MDR-9050210. TERC is a nonprofit company working to improve mathematics and science education. TERC is located at 2067 Massachusetts Avenue, Cambridge, MA 02140.

This project was supported, in part, by the
National Science Foundation
Opinions expressed are those of the authors and not necessarily those of the Foundation

This book is published by Dale Seymour Publications, an imprint of the Alternative Publishing Group of Addison-Wesley Publishing Company.

Project Editor: Priscilla Cox-Samii
Series Editor: Beverly Cory
Manuscript Editor: Sandra Sella Raas
ESL Consultant: Nancy Sokol Green
Production/Manufacturing Director: Janet Yearian
Production/Manufacturing Coordinator: Barbara Atmore
Design Manager: Jeff Kelly
Design: Don Taka
Illustrations: DJ Simison, Carl Yoshihara
Composition: Publishing Support Services
Cover: Bay Graphics

Copyright © 1995 by Dale Seymour Publications. All rights reserved.
Printed in the United States of America.

 Printed on Recycled Paper

Limited reproduction permission: The publisher grants permission to individual teachers who have purchased this book to reproduce the blackline masters as needed for use with their own students. Reproduction for an entire school or school district or for commercial use is prohibited.

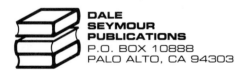

DALE SEYMOUR PUBLICATIONS
P.O. BOX 10888
PALO ALTO, CA 94303

Order number DS21239
ISBN 0-86651-801-0
2 3 4 5 6 7 8 9 10-ML-98 97 96 95 94

T E R C

INVESTIGATIONS IN NUMBER, DATA, AND SPACE

Principal Investigator Susan Jo Russell
Co-Principal Investigator Cornelia C. Tierney
Director of Research and Evaluation Jan Mokros

Curriculum Development
Joan Akers
Michael T. Battista
Mary Berle-Carman
Douglas H. Clements
Karen Economopoulos
Ricardo Nemirovsky
Andee Rubin
Susan Jo Russell
Cornelia C. Tierney
Amy Shulman Weinberg

Evaluation and Assessment
Mary Berle-Carman
Abouali Farmanfarmaian
Jan Mokros
Mark Ogonowski
Amy Shulman Weinberg
Tracey Wright
Lisa Yaffee

Teacher Development and Support
Rebecca B. Corwin
Karen Economopoulos
Tracey Wright
Lisa Yaffee

Technology Development
Michael T. Battista
Douglas H. Clements
Julie Sarama Meredith
Andee Rubin

Video Production
David A. Smith

Administration and Production
Amy Catlin
Amy Taber

Cooperating Classrooms for This Unit
Katie Bloomfield
Robert A. Dihlmann
Shutesbury Elementary, Shutesbury, MA

Corrine Varon
Virginia M. Micciche
Cambridge Public Schools, Cambridge, MA

Joan Forsyth
Jeanne Wall
Arlington Public Schools, Arlington, MA

Consultants and Advisors
Elizabeth Badger
Deborah Lowenberg Ball
Marilyn Burns
Ann Grady
Joanne M. Gurry
James J. Kaput
Steven Leinwand
Mary M. Lindquist
David S. Moore
John Olive
Leslie P. Steffe
Peter Sullivan
Grayson Wheatley
Virginia Woolley
Anne Zarinnia

Graduate Assistants
Kent State University:
Joanne Caniglia, Pam DeLong, Carol King

State University of New York at Buffalo:
Rosa Gonzalez, Sue McMillen,
Julie Sarama Meredith, Sudha Swaminathan

CONTENTS

About the *Investigations* Curriculum   1
How to Use This Book   2
About Assessment   7

Things That Come in Groups
    Overview   8
    Materials List   11
    About the Mathematics in This Unit   12
    Preview for the Linguistically Diverse Classroom   13

Investigation 1: Things That Come in Groups   16
    Session 1: Many Things Come in Groups   18
    Session 2: How Many in Several Groups?   21
    Session 3: Writing and Solving Riddles   26
    Session 4: (Excursion): Each Orange Had 8 Slices   30

Investigation 2: Skip Counting and 100 Charts   34
    Session 1: Highlighting Multiples in 100 Charts   36
    Session 2: Using the Calculator to Skip Count   40
    Sessions 3 and 4: More Practice with Multiples   42
    Sessions 5 and 6: Discussing Number Patterns   47

Investigation 3: Arrays and Skip Counting   54
    Session 1: Arranging Chairs   56
    Session 2: Array Games   62
    Session 3: The Shapes of Arrays   67

Investigation 4: The Language of Multiplication and Division   70
    Sessions 1 and 2: Multiply or Divide?   72
    Sessions 3 and 4: Writing and Solving Story Problems   84

Investigation 5: Problems with Larger Numbers   88
    Session 1: Calculating Savings   90
    Session 2: Many, Many Legs   93
    Session 3: Data Tables and Line Plots   97
    Session 4: A Riddle with 22 Legs   101

Appendix: Ten-Minute Math   105
Appendix: Vocabulary Support for Second-Language Learners   109
Blackline Masters   111
    Family Letter
    Student Sheets 1–8
    Teaching Resources

Teacher Notes

What About Notation?	25
The Relationship Between Division and Multiplication	29
Students' Problems with Skip Counting	39
Introducing Mathematical Vocabulary	46
Assessment: Arrays That Total 36	69
Talking and Writing About Division	81
Two Kinds of Division: Sharing and Partitioning	82

ABOUT THE *INVESTIGATIONS* CURRICULUM

Investigations in Number, Data, and Space is a K–5 mathematics curriculum with four major goals:

- to offer students meaningful mathematical problems
- to emphasize depth in mathematical thinking rather than superficial exposure to a series of fragmented topics
- to communicate mathematics content and pedagogy to teachers
- to substantially expand the pool of mathematically literate students

The *Investigations* curriculum embodies an approach radically different from the traditional textbook-based curriculum. At each grade level, it consists of a set of units, each offering 2–4 weeks of work. These units of study are presented through investigations that involve students in the exploration of major mathematical ideas.

Approaching the mathematics content through investigations helps students develop flexibility and confidence in approaching problems, fluency in using mathematical skills and tools to solve problems, and proficiency in evaluating their solutions. Students also build a repertoire of ways to communicate about their mathematical thinking, while their enjoyment and appreciation of mathematics grows.

The investigations are carefully designed to invite all students into mathematics—girls and boys, diverse cultural, ethnic, and language groups, and students with different strengths and interests. Problem contexts often call on students to share experiences from their family, culture, or community. The curriculum eliminates barriers—such as work in isolation from peers, or emphasis on speed and memorization—that exclude some students from participating successfully in mathematics. The following aspects of the curriculum ensure that all students are included in significant mathematics learning:

- Students spend time exploring problems in depth.
- They find more than one solution to many of the problems they work on.
- They invent their own strategies and approaches, rather than relying on memorized procedures.
- They choose from a variety of concrete materials and appropriate technology, including calculators, as a natural part of their everyday mathematical work.
- They express their mathematical thinking through drawing, writing, and talking.
- They work in a variety of groupings—as a whole class, individually, in pairs, and in small groups.
- They move around the classroom as they explore the mathematics in their environment and talk with their peers.

While reading and other language activities are typically given a great deal of time and emphasis in elementary classrooms, mathematics often does not get the time it needs. If students are to experience mathematics in depth, they must have enough time to become engaged in real mathematical problems. We believe that a minimum of five hours of mathematics classroom time a week—about an hour a day—is critical at the elementary level. The plan and pacing of the *Investigations* curriculum is based on that belief.

For further information about the pedagogy and principles that underlie these investigations, see the Teacher Notes throughout the units and the following books:

- *Implementing the* Investigations in Number, Data, and Space™ *Curriculum*
- *Beyond Arithmetic*

HOW TO USE THIS BOOK

The *Investigations* curriculum is presented through a series of teacher books, one for each unit of study. These books not only provide a complete mathematics curriculum for your students, they offer materials to support your own professional development. You, the teacher, are the person who will make this curriculum come alive in the classroom; the book for each unit is your main support system.

While reproducible resources for students are provided, the curriculum does not include student books. Students work actively with objects and experiences in their own environment and with a variety of manipulative materials and technology, rather than with workbooks and worksheets filled with problems. We also make extensive use of the overhead projector as a way to present problems, to focus group discussion, and to help students share ideas and strategies. If an overhead projector is available, we urge you to try it as suggested in the investigations.

Ultimately, every teacher will use these investigations in ways that make sense for his or her particular style, the particular group of students, and the constraints and supports of a particular school environment. We have tried to provide with each unit the best information and guidance for a wide variety of situations, drawn from our collaborations with many teachers and students over many years. Our goal in this book is to help you, as a professional educator, implement this mathematics curriculum in a way that will give all your students access to mathematical power.

Investigation Format

The opening two pages of each investigation help you get ready for the student work that follows. Here you will read:

What Happens—a synopsis of each session or block of sessions.

Mathematical Emphasis—the most important ideas and processes students will encounter in this investigation.

What to Plan Ahead of Time—materials to gather, student sheets to duplicate, transparencies to make, and anything else you need to do before starting.

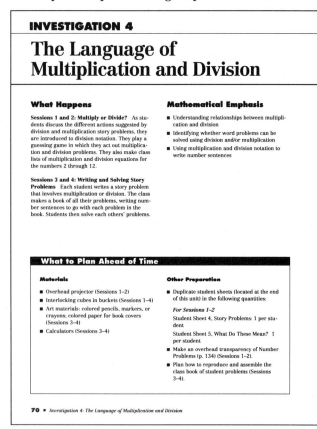

2 ■ *Things That Come in Groups*

Sessions Within an investigation, the activities are organized by class session, a session being a one-hour math class. Sessions are numbered consecutively through an investigation. Often several sessions are grouped together, presenting a block of activities with a single major focus.

When you find a block of sessions presented together—for example, Sessions 1, 2, and 3—read through the entire block first to understand the overall flow and sequence of the activities. Make some preliminary decisions about how you will divide the activities into three sessions for your class, based on what you know about your students. You may need to modify your initial plans as you progress through the activities, and you may want to make notes in the margins of the pages as reminders for the next time you use the unit.

Be sure to read the Session Follow-Up section at the end of the session block to see what homework assignments and extensions are suggested as you make your initial plans.

While you may be used to a curriculum that tells you exactly what each class session should cover, we have found that the teacher is in a better position to make these decisions. Each unit is flexible and may be handled somewhat differently by every teacher. While we provide guidance for how many sessions a particular goup of activities is likely to need, we want you to be active in determining an appropriate pace and the best transition points for your class.

Ten-Minute Math At the beginning of some sessions, you will find Ten-Minute Math activities. These are designed to be used in tandem with the investigations, but not during the math hour. Rather, we hope you will do them whenever you have a spare 10 minutes—maybe before lunch or recess, or at the end of the day.

Ten-Minute Math offers practice in key concepts, but not always those being covered in the unit. For example, in a unit on using data, Ten-Minute Math might revisit geometric activities done earlier in the year. Complete directions for the suggested activities are included at the end of each unit. A compilation of Ten-Minute Math activities is also available as a separate book.

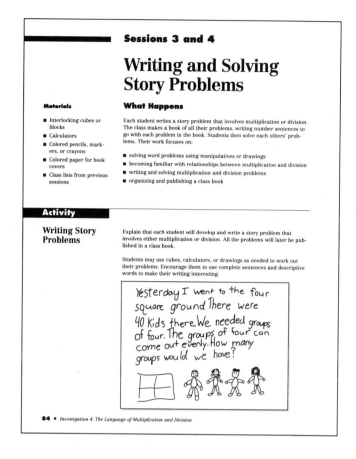

Activities The activities include pair and small-group work, individual tasks, and whole-class discussions. In any case, students are seated together, talking and sharing ideas during all work times. Students most often work cooperatively, although each student may record work individually.

Choice Time Some sessions are structured with activity choices. In these cases, students may work simultaneously on different activities focused on the same mathematical ideas. Students choose which activities they want to do, and they cycle through them.

You will need to decide how to set up and introduce these activities and how to let students make their choices. Some teachers present them as station activities, in different parts of the room. Some list the choices on the board as reminders or have students keep their own lists.

Excursions Some of the investigations in this unit include *excursions*—sessions that could be omitted without harming the integrity of the unit. This is one way of dealing with the overabundance of fascinating mathematics to be studied—much more than a class has time to explore in any one year. Excursions give you the flexibility to make different choices from year to year. For example, you might do the excursions in this Multiplication and Division unit this year, but another year, try the excursions in another unit.

Tips for the Linguistically Diverse Classroom
At strategic points in each unit, you will find concrete suggestions for simple modifications of the teaching strategies to encourage the participation of all students. Many of these tips offer alternative ways to elicit critical thinking from students at varying levels of English proficiency, as well as from other students who find it difficult to verbalize their thinking.

The tips are supported by suggestions for specific vocabulary work to help ensure that all students can participate fully in the investigations. The Preview for the Linguistically Diverse Classroom (p.13) lists important words that are assumed as part of the working vocabulary of the unit. Second-language learners will need to become familiar with these words in order to understand the problems and activities they will be doing. These terms can be incorporated into students' second-language work before or during the unit. Activities that can be used to present the words are found in the appendix, Vocabulary Support for Second-Language Learners (p. 109).

In addition, ideas for making connections to students' language and cultures, included on the Preview page, help the class explore the unit's concepts from a multicultural perspective.

Session Follow-Up

Homework Homework is not given daily for its own sake, but periodically as it makes sense to have follow-up work at home. Homework may be used for (1) review and practice of work done in class; (2) preparation for activities coming up—for example, collecting data for a class project; or (3) involving and informing family members.

Some units in the *Investigations* curriculum have more homework than others, simply because it makes sense for the mathematics that's going on. Other units rely on manipulatives that most students won't have at home, making homework diffi-

cult. In any case, homework should always be directly connected to the investigations in the unit, or to work in previous units—never sheets of problems just to keep students busy.

Extensions These follow-up activities are opportunities for some or all students to explore a topic in greater depth or in a different context. They are not designed only for "fast" students; mathematics is a multifaceted discipline, and different students will want to go further in different investigations. Look for and encourage the sparks of interest and enthusiasm you see in your students, and use the extensions to help them pursue these interests.

Family Letter A letter that you can send home to students' families is included with the blackline masters for each unit. We want families to be informed about the mathematics work in your classroom; they should be encouraged to participate in and support their children's work. A reminder to send home the letter appears in one of the early investigations. (These letters are also available separately in the following languages: Spanish, Vietnamese, Cantonese, Hmong, and Cambodian.)

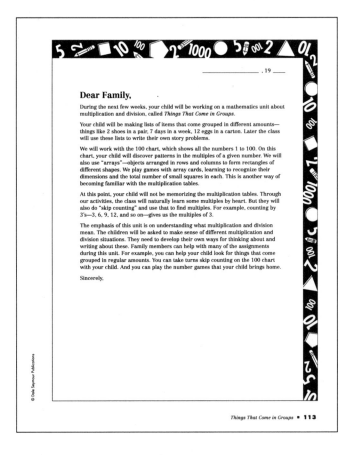

Materials

A complete list of the materials needed for the unit is found on p. 11. Some of these materials are available in a kit for the Investigations grade 3 curriculum. Individual items can also be purchased as needed from school supply stores and dealers.

In an active mathematics classroom, certain basic materials should be available at all times: interlocking cubes, pencils, unlined paper, graph paper, calculators, things to count with, and measuring tools. Some activities in this curriculum require scissors and glue sticks or tape. Stick-on notes and large paper are also useful materials throughout.

So that students can independently get what they need at any time, they should know where these materials are kept, how they are stored, and how they are to be returned to the storage area. For example, interlocking cubes are best stored in towers of ten; then, whatever the activity, they should be returned to storage in groups of ten at the end of the hour. You'll find that establishing such routines at the beginning of the year is well worth the time and effort.

Student Sheets and Teaching Resources

Reproducible pages to help you teach the unit are found at the end of this book. These include masters for making overhead transparencies and other teaching tools, as well as student recording sheets.

Many of the field-test teachers requested more sheets to help students record their work, and we have tried to be responsive to this need. At the same time, we think it's important that students find their own ways of organizing and recording their work. They need to learn how to explain their thinking with both drawings and written words, and how to organize their results so someone else can understand them.

To ensure that students get a chance to learn how to represent and organize their own work, we deliberately do not provide student sheets for every activity. We recommend that your students keep a mathematics notebook or folder so that their work, whether on reproducible sheets or their own paper, is always available to them for reference.

Help for You, the Teacher

Because we believe strongly that a new curriculum must help teachers think in new ways about mathematics and about their students' mathematical thinking processes, we have included a great deal of material to help you learn more about both.

About the Mathematics in This Unit This introductory section (p. 12) summarizes for you the critical information about the mathematics you will be teaching. This will be particularly valuable to teachers who are accustomed to a traditional textbook-based curriculum.

Teacher Notes These reference notes provide practical information about the mathematics you are teaching and about our experience with how students learn. Many of the notes were written in response to actual questions from teachers, or to discuss important things we saw happening in the field-test classrooms. Some teachers like to read them all before starting the unit, then review them as they come up in particular investigations.

Dialogue Boxes Sample dialogues throughout the unit demonstrate how students typically express their mathematical ideas, what issues and confusions arise in their thinking, and how some teachers have guided class discussions.

These dialogues are based on the extensive classroom testing of this curriculum; many are word-for-word transcriptions of recorded class discussions. They are not always easy reading; sometimes it may take some effort to unravel what the students are trying to say. But this is the value of these dialogues; they offer good clues to how your students may develop and express their approaches and strategies, helping you prepare for your own class discussions.

Where to Start You may not have time to read everything the first time you use this unit. As a first-time user, you will likely focus on understanding the activities and working them out with your students. Read completely through each investigation before starting to present it.

When you next teach this same unit, you can begin to read more of the background. Each time you present this unit, you will learn more about how your students understand the mathematical ideas. The first-time user of *Things That Come in Groups* should read the following:

- About the Mathematics in This Unit (p. 12)
- Teacher Note: What About Notation? (p. 25)
- Teacher Note: The Relationship Between Division and Multiplication (p. 29)
- Teacher Note: Talking and Writing About Division (p. 81)

ABOUT ASSESSMENT

Teacher Checkpoints As a teacher of the *Investigations* curriculum, you observe students daily, listen to their discussions, look carefully at their work, and use this information to guide your teaching. We have designated Teacher Checkpoints as natural times to get an overall sense of how your class is doing in the unit.

The Teacher Checkpoints provide a time for you to pause and reflect on your teaching plan while observing students at work in an activity. These sections offer tips on what you should be looking for and how you might adjust your pacing. Are most students fluent with strategies for solving a particular kind of problem? Are they just starting to formulate good strategies? Or are they still struggling with how to start?

Depending on what you see as the students work, you may want to spend more time on similar problems, change some of the problems to use smaller numbers, move quickly to more challenging material, modify subsequent activities for some students, work on particular ideas with a small group, or pair students who have good strategies with those who are having more difficulty.

In this unit you will find three Teacher Checkpoints:

> Do They Understand Multiplication? (p. 28)
> Using the Skip Counting Circles (p. 45)
> Do They Understand the Notation? (p. 80)

Embedded Assessment Activities Use the built-in assessments included in this unit to help you examine the work of individual students, figure out what it means, and provide feedback. From the students' point of view, the activities you will be using for assessment are no different from any others; they don't look or feel like traditional tests.

These activities sometimes involve writing and reflecting, at other times a brief interaction between student and teacher, and in still other instances the creation and explanation of a product.

In this unit you will find assessment activities in the third and fifth investigations:

> Arrays That Total 36 (p. 68)
> A Riddle with 22 Legs (p. 101)

Teachers find the hardest part of the assessment to be interpreting their students' work. If you have used a process approach to teaching writing, you will find our mathematics approach familiar. To help with interpretation, we provide guidelines and questions to ask about the students' work. In some cases we include a Teacher Note with specific examples of student work and a commentary on what it indicates. This framework can help you determine how your students are progressing.

As you evaluate students' work, it's important to remember that you're looking for much more than the "right answer." You'll want to know what their strategies are for solving the problem, how well these strategies work, whether they can keep track of and logically organize an approach to the problem, and how they make use of representations and tools to solve the problem.

Ongoing Assessment Good assessment of student work involves a combination of approaches. Some of the things you might do on an ongoing basis include the following:

- **Observation** Circulate around the room to observe students as they work. Watch for the development of their mathematical strategies, and listen to their discussions of mathematical ideas.

- **Portfolios** Ask students to document their work in journals, notebooks, or portfolios. Periodically review this work to see how their mathematical thinking and writing are changing. Some teachers have students keep a notebook or folder for each unit, while others prefer one mathematics notebook or a portfolio of selected work for the entire year. Take time at the end of each unit for students to choose work for their portfolios. You might also have them write about what they've learned in the unit.

Things That Come in Groups

OVERVIEW

Content of This Unit To develop experience with some uses of multiplication and division, students work with things that come in groups, with patterns in the multiplication tables using 100 charts, and with rectangular arrays. They invent and solve problems about the number of legs on living creatures. Students become familiar with the multiplication tables up to the 12's, with emphasis on multiples with totals under 50. They also invent their own ways of solving multiplication and division problems.

Connections with Other Units If you are doing the full-year *Investigations* curriculum in the suggested sequence for grade 3, this is the second of 10 units. This unit introduces a variety of activities that use small numbers; these activities will be repeated later in *Landmarks in the Hundreds*, using larger numbers—with factors and multiples of 100.

This unit can be used successfully at either grade 3 or grade 4 as an introduction to multiplication and division, depending on the previous experience and needs of your students. Many of the ideas and games in this unit are picked up and extended in the grade 4 Multiplication and Division units.

Investigations Curriculum ■ Suggested Grade 3 Sequence

Mathematical Thinking at Grade 3 (Introduction)

▶ *Things That Come in Groups* (Multiplication and Division)

Flips, Turns, and Area (2-D Geometry)

From Paces to Feet (Measuring and Data)

Landmarks in the Hundreds (The Number System)

Up and Down the Number Line (Changes)

Combining and Comparing (Addition and Subtraction)

Turtle Paths (2-D Geometry)

Fair Shares (Fractions)

Exploring Solids and Boxes (3-D Geometry)

Investigation 1 • Things That Come in Groups				
Class Sessions	Activities	Pacing	Ten-Minute Math
Session 1 MANY THINGS COME IN GROUPS	Naming Things That Come in Groups Asking Multiplication Questions Brainstorming About Groups ■ Homework	1 hr	
Session 2 HOW MANY IN SEVERAL GROUPS?	Pictures of Things That Come in Groups Writing "Groups of" as Multiplication ■ Homework	1 hr	
Session 3 WRITING AND SOLVING RIDDLES	Writing Riddles for Our Pictures ■ Teacher Checkpoint: Do They Understand Multiplication? ■ Extension	1 hr	Counting Around the Class
Session 4 (Excursion)* EACH ORANGE HAD 8 SLICES	How Many Altogether?	1 hr	

Investigation 2 • Skip Counting and 100 Charts				
Class Sessions	Activities	Pacing	Ten-Minute Math
Session 1 HIGHLIGHTING MULTIPLES IN 100 CHARTS	Highlighting 2's and 3's Making Books of 100 Charts ■ Homework	1 hr	
Session 2 USING THE CALCULATOR TO SKIP COUNT	Skip Counting 4's and More ■ Homework	1 hr	
Sessions 3 and 4 MORE PRACTICE WITH MULTIPLES	Choice Time: Exploring Multiples and Patterns ■ Teacher Checkpoint: Using the Skip Counting Circles ■ Homework	2 hrs	Counting Around the Class
Sessions 5 and 6 DISCUSSING NUMBER PATTERNS	Patterns in Multiples of 9 and 11 Numbers That Appear on Two Charts Discussion: Patterns Across the Charts Playing Cover 50 ■ Homework	2 hrs	

*Excursions can be omitted without harming the integrity or continuity of the unit, but offer good mathematical work if you have time to include them.

Continued on next page

Unit Overview ■ 9

Investigation 3 • Arrays and Skip Counting

Class Sessions	Activities	Pacing	Ten-Minute Math
Session 1 ARRANGING CHAIRS	Arranging Chairs in Rectangular Arrays Arranging More Chairs Making Array Cards ■ Homework	1 hr	
Session 2 ARRAY GAMES	Counting Squares in Arrays Playing Array Games ■ Homework ■ Extension	1 hr	Counting Around the Class
Session 3 THE SHAPES OF ARRAYS	Discussing Array Game Strategies ■ Assessment: Arrays That Total 36 ■ Homework	1 hr	

Investigation 4 • The Language of Multiplication and Division

Class Sessions	Activities	Pacing	Ten-Minute Math
Sessions 1 and 2 MULTIPLY OR DIVIDE?	Solving Story Problems Acting Out Number Sentences Different Ways to Write Problems Writing Multiplication and Division Sentences ■ Teacher Checkpoint: Do They Understand the Notation?	2 hrs	
Sessions 3 and 4 WRITING AND SOLVING STORY PROBLEMS	Writing Story Problems A Class Book of Problems Solving Problems in the Class Book ■ Homework ■ Extension	2 hrs	Likely or Unlikely?

Investigation 5 • Problems with Larger Numbers

Class Sessions	Activities	Pacing	Ten-Minute Math
Session 1 CALCULATING SAVINGS	How Much Would You Save? ■ Homework ■ Extension	1 hr	
Session 2 MANY, MANY LEGS	Discussion: What Could We Buy? How Many Legs? Planning a Survey ■ Homework	1 hr	Likely or Unlikely?
Session 3 DATA TABLES AND LINE PLOTS	Expanding Our Data Tables Making a Line Plot Problems from Our Own Data ■ Homework	1 hr	
Session 4 A RIDDLE WITH 22 LEGS	■ Assessment: A Riddle with 22 Legs	1 hr	

Things That Come in Groups

MATERIALS LIST

Following are the basic materials needed for the activities in this unit. The suggested quantities are ideal; however, in some instances you can work with smaller quantities by running several activities, requiring different materials, simultaneously.

Items marked with an asterisk are available in the *Investigations* Materials Kit for grade 3.

* Interlocking cubes: 50 per student
* Array Cards (manufactured; or use blackline masters to make your own sets)

　Calculators: at least 1 per pair of students

　Each Orange Had 8 Slices by Paul Giganti, Jr. and Donald Crews (Greenwillow, 1992) (optional)

　Scissors: 1 per student

　Legal-size envelopes

　Quart-size resealable plastic bags

　Large paper for making class lists

　Colored paper, drawing paper

　Colored pencils, markers, or crayons

　Overhead projector

　Blank overhead transparencies, pens

The following materials are provided at the end of this unit as blackline masters. They are also available in classroom sets.

Family letter (p. 113)

Student Sheets 1–8 (pp. 114–121)

Teaching Resources:

　Cover 50 Game (p. 122)

　How to Play Cover 50 (p. 123)

　How to Make Array Cards (p. 124)

　Array Cards, Sheets 1–6 (pp. 125–130)

　The Arranging Chairs Puzzle (p. 131)

　How to Play Multiplication Pairs (p. 132)

　How to Play Count and Compare (p. 133)

　Number Problems (transparency master) (p. 134)

　Half-inch graph paper (p. 135)

ABOUT THE MATHEMATICS IN THIS UNIT

In this unit, students develop their own strategies for doing multiplication and division problems. They discover that both types of problems deal with equal groups, but each will answer different questions about the groups. Multiplication is typically used when the size of each group and the number of groups is known, and we want to find the total number of items. Division is most often used when the total quantity is known, and we want to find out either the number or the size of the groups.

As students develop strategies to use in multiplication and division situations, it is critical that they develop visual images that support their work. They may use an array or squares, for example, to visualize an important multiplication relationship—that the solution to 7×6 is the same as the solution to 6×7. As students skip count on a 100 chart, they begin to recognize characteristics of particular multiples. They will see, for example, that all the multiples of 2, 4, and 6 are even numbers, or that all the multiples of 5 end in either 5 or 0. Students may at first visualize multiplication as repeated addition, since this process is more familiar to them.

Throughout this unit, it is most important to support students' efforts to make sense out of multiplication and division. As students develop their own strategies, they are aided by knowing many of the single-digit multiplication pairs. We do not expect them to memorize all the multiples, but as they look at patterns in the tables and construct the multiples again and again by skip counting, students will commit many of them to memory. They will also pick up ways to solve others quickly—for example, by using a known answer to find an unknown one ("8×6 is like 4×6 twice, so it's 24 and 24, and that's 48").

Students also learn to read standard multiplication and division notation and to use this notation to record their work. They must also learn that notation communicates the problem to be solved, but doesn't prescribe the method of solution.

When students see problems written in standard forms such as these:

$$\begin{array}{r} 56 \\ \times 8 \end{array} \qquad 4\overline{)132}$$

the form of the problem may trigger use of poorly understood, and often inefficient, algorithms. For example, in the first problem, students might start to say, "8 times 6 is 48, put down the 8 and carry the 4...." This procedure obscures the use of good number sense and often leads students to fragment a number into its digits and lose track of the quantities represented by the numerals. Good mental strategies often start from the left, focusing first on the largest part of the number, rather than the smallest: "eight 50's is 400, eight 6's is 48, so that's 448."

Students need to develop efficient computation strategies, many of which will be mental strategies, but these must be based on their understanding of the quantities and their relationships, not on memorized procedures. We would like students to recognize multiplication and division problems written in all of the notations they are likely to see in elementary school, but to solve them in their own way.

Mathematical Emphasis At the beginning of each investigation, the Mathematical Emphasis section tells you what is most important for students to learn about during that investigation. Many of these mathematical understandings are difficult and complex. Students gradually learn more and more about each idea over many years of schooling. Individual students will begin and end the unit with different levels of knowledge and skill, but all will gain greater knowledge about multiples and factors and about some meanings and notation for multiplication and division.

PREVIEW FOR THE LINGUISTICALLY DIVERSE CLASSROOM

In the *Investigations* curriculum, mathematical vocabulary is introduced naturally during the activities. We don't ask students to learn definitions of new terms; rather, they come to understand such words as *factor* or *area* or *symmetry* by hearing them used frequently in discussion as they investigate new concepts. This approach is compatible with current theories of second-language acquisition, which emphasize the use of new vocabulary in meaningful contexts while students are actively involved with objects, pictures, and physical movement.

Listed below are some key words used in this unit that will not be new to most English speakers at this age level, but may be unfamiliar to students with limited English proficiency. You will want to spend additional time working on these words with your students who are learning English. If your students are working with a second-language teacher, you might enlist your colleague's aid in familiarizing students with these words, before and during this unit. In the classroom, look for opportunities for students to hear and use these words. Activities you can use to present the words are given in the appendix, Vocabulary Support for Second-Language Learners (p. 109).

question, statement, illustrate   In Investigation 1, students write and *illustrate* their own multiplication "riddles," making two or more *statements* involving numbers, and ending with a *question*.

chart, row, column   Students color in multiples on the 100 *chart* in Investigation 2, looking for visual patterns in its *rows* and *columns*, as well as diagonals.

calculator, press, equals key, plus key   Students learn to use the calculator to skip count by any number, *pressing* that number, the *plus key*, and then the *equals key* repeatedly.

day, week, month, amount   In a savings problem in Investigation 5, students find the *amount* of money they would save in a *week* or a *month* by saving the same amount each *day*.

creature, leg, -legged   In a series of activities during Investigation 5 that involve working with larger numbers, students investigate various combinations of creatures with different numbers of legs.

Multicultural Extensions for All Students

Whenever possible, encourage students to share words, objects, customs, or any aspects of daily life from their own cultures and backgrounds that are relevant to the activities in this unit. For example:

- When students are thinking of things that come in groups during the first investigation, encourage them to include groups of things that may reflect their culture—such as the number of dancers in a particular dance, or the number of playing pieces in a popular game.

- When students are writing story problems in Investigation 4, encourage the use of culture-specific references (to items of food, for example) so that in sharing their problems with the class, they share a little of themselves as well.

Investigations

INVESTIGATION 1

Things That Come in Groups

What Happens

Session 1: Many Things Come in Groups As a class, students make lists of things that come in groups of 2 to 12. They figure out how many things come in two groups of 2, three groups of 2, and so on, in order to create multiplication problems based on the class lists. Partners then trade problems for each other to solve.

Session 2: How Many in Several Groups? Students choose a multiplication situation to illustrate. They make an illustration and then describe it in words. They are also introduced to number sentences that describe multiplication situations.

Session 3: Writing and Solving Riddles Students pose problems about their pictures by leaving out one of the two factors or the answer, then solve each other's riddles.

Session 4 (Excursion): Each Orange Had 8 Slices Students listen to the problems presented in the book *Each Orange Had 8 Slices* by Paul Giganti, Jr., and Donald Crews. Students track the two- and three-step problems and try to solve them on their own.

Mathematical Emphasis

- Finding things that come in groups
- Using multiplication notation
- Using multiplication to mean groups of groups
- Writing and illustrating multiplication sentences

What to Plan Ahead of Time

Materials

- Large paper: about 15 sheets (Session 1)
- Art materials: paper; colored pencils, markers, or crayons (Session 2)
- Interlocking cubes or blocks: 50 per student (Sessions 3–4)
- *Each Orange Had 8 Slices* by Paul Giganti, Jr., and Donald Crews (Greenwillow, 1992) (Session 4, optional)
- Calculators (Session 4)

Other Preparation

- Duplicate the family letter (p. 113) to send home after Session 1. Remember to sign it.
- If you plan to provide folders in which students will save their work for the entire unit, prepare these for distribution during Session 1.

INVESTIGATION 1

Investigation 1: Things That Come in Groups ■ 17

Session 1

Many Things Come in Groups

Materials

- Large paper (for class lists)

What Happens

As a class, students make lists of things that come in groups of 2 to 12. Students figure out how many things come in two groups of 2, three groups of 2, and so on in order to create multiplication problems based on the class lists. Partners then trade problems for each other to solve. Student work focuses on:

- finding things that come in groups of certain sizes
- recognizing multiplication situations
- posing and solving multiplication problems

Activity

Naming Things That Come in Groups

Briefly introduce this activity and provide examples to prompt thinking:

Many things come in groups. Some products are always packaged with the same number of items. Art and office supplies (pencils, paper clips) and some foods (frankfurters, eggs) often come in groups. Many things in nature (petals on a flower, toes on a cat) come in certain numbers. What things can *you* think of that come in groups?

Start a list of things on separate pieces of large paper for each number from 2 to 12, and for any other numbers that students mention two or more times; 24, 60, or 100, for example. Under each number, record a few of the students' ideas.

2	3	4	5
eyes	juice boxes	legs on a dog	fingers on one hand
ears	clover leaves	seasons in a year	toes on one foot
twins		quarters in a dollar	school days in a week
shoes			cents in a nickel
mittens			senses

18 ■ *Investigation 1: Things That Come in Groups*

On each list, leave space to add new ideas or examples. Also leave room for posting a list of multiplication equations for that table, which you will be writing as a group in Investigation 4. Post the lists in numerical order where students can see them, perhaps temporarily on the chalkboard. Save the lists for use in the next session.

❖ **Tip for the Linguistically Diverse Classroom**   In situations like this throughout the unit, add small drawings for the listed items as needed to enhance comprehension. Students with limited English proficiency can contribute ideas by adding their own small sketches to the chart.

Activity

Asking Multiplication Questions

To help students become familiar with situations where multiplication could be used, pose a few questions that can be solved using multiplication. Base your questions on items from the class lists. For example:

We agree that there are usually five toes on a person's foot. How many toes would there usually be on four feet?

There are four quarters in a dollar. How many quarters are in three dollars?

Writing and Solving Multiplication Questions   After students have listened to and answered a few of your questions, ask them to make up a question of their own. They can use items from the posted lists or think of new ones.

Every student should write a multiplication question like the examples you just gave. When they are finished, have them trade questions with a partner. Partners then copy each other's problem onto their paper and write solutions to both their own and their partner's problems.

❖ **Tip for the Linguistically Diverse Classroom**   Have students include drawings for the items in their problems as an aid to comprehension.

Partners then check each other's work. If they want, students may use cubes or other small objects as counters.

After students have had time to develop and solve their multiplication questions alone and with a partner, bring the class back together. Invite a few volunteers to read their questions aloud for the class to solve.

Many Things Come in Groups ■ **19**

Activity

Brainstorming About Groups

Working with partners or in small groups, students spend the rest of the class time brainstorming more things that come in groups. Encourage them to think of things in groups of 2 to 12. They may use larger numbers only if they can think of more than one item in each grouping. Provide some time for students to add the new ideas to the class lists.

When students have added their new ideas, bring them together to talk about any observations they might have about their lists. Ask questions such as these:

What are the most common numbers that things are grouped in? Which numbers were hard to find? Why do you think some numbers occur more than others?

Challenge students to try to think of something for every one of the numbers from 2 to 12 over the next few days.

Note: Be sure to save the class lists of things that come in groups for use in the next session.

Session 1 Follow-Up

Homework

Send home the family letter. For homework, students talk with people at home about things that come in groups. They list some new things they think of or find, either at home or at a store, writing down the name and the quantity of each item. If it is a packaged item, they might include the brand name. Encourage them to try to find items in different amounts, especially some of the hard-to-find numbers, such as things that come in groups of 7 or 11.

❖ **Tip for the Linguistically Diverse Classroom** Students may list items in their native languages and draw pictures of the things they find that come in groups.

Session 2

How Many in Several Groups?

What Happens

Students choose a multiplication situation to illustrate. They make an illustration and then describe it in words. They are introduced to number sentences that describe multiplication situations. Their work focuses on:

- illustrating a multiplication situation
- describing multiplication situations in words and numbers

Materials

- Class lists of things that come in groups (from Session 1)
- Colored pencils, markers, or crayons

 Ten-Minute Math: Counting Around the Class Once or twice during the next few days, do Counting Around the Class. Remember that Ten-Minute Math activities are best done outside of math time.

Some students feel uncomfortable with this activity unless they are prepared to participate. For suggestions about how to maximize all students' participation and confidence, see the full description of Counting Around the Class on p. 105.

Choose a number to count by, let's say 2. Ask students to predict what number they'll land on if they count around the class exactly once by that number. Encourage students to talk about how they could figure this out without doing the actual counting.

Then start the count: the first student says "2," the next "4," the next "6," and so forth. If students seem unfamiliar with what comes next, you may want to put numbers on the board as they count, so they can begin to see patterns.

Stop two or three times during the count to ask questions like this:

We're at 16—how many students have counted so far?

After counting around once, compare the actual ending number with their predictions.

Activity

Pictures of Things That Come in Groups

Take a few minutes to add to the posted class lists some new items that students discovered for homework. Focus on the number lists where there are only a few items so far. If necessary, give students extra time later in the day to add more of their ideas to the lists.

Referring to the posted lists of things that come in groups, each student chooses an item and illustrates a few groups of that item. For example, a student might show two basketball teams as two groups of 5, or three cartons of eggs as three groups of 12.

To ensure that many different combinations are illustrated by the class as a whole, you might assign numbers in some way. One possibility is to assign each number to a different pair of students. Together, they can make several different pictures of groups of that size, each picture showing a different number of groups. For example, a student pair assigned the number 3 might draw 4 storybooks of "The Three Bears," 2 people making 3 wishes each, and 5 three-leaf clovers. Suggest that students start with no more than 6 groups. They can illustrate more groups later.

Students must draw their pictures clearly enough for the items in each group to be counted. Then they write brief sentences describing the groups and the total number of items represented in their picture.

❖ **Tip for the Linguistically Diverse Classroom** Pair students who are proficient in English with any who are not yet writing in English, to help write the sentences that describe the illustrated groups of things, perhaps leaving blanks for the numbers to be filled in by the student who made the drawing.

Activity

Writing "Groups of" as Multiplication

Use the students' pictures as an opportunity to introduce symbolic notation for multiplication.

We can describe the problems on your pictures using addition. For example, 6 cars have 4 + 4 + 4 + 4 + 4 + 4 = 24 wheels. [*Write this addition problem on the board.*] **We can also write 6 groups of 4 as a multiplication sentence: 6 × 4 = 24.** [*Write this multiplication notation next to it.*]

Ask a few volunteers to share their pictures. Each student shows a picture and reads aloud the sentences that describe it. Other students can then tell how to write the number sentence for the problem. Write some of their examples on the board:

2 cartons of 12 eggs are 24 eggs altogether.	2 × 12 = 24
4 cars with 4 wheels each have 16 wheels in all.	4 × 4 = 16

Students decide what the multiplication number sentence is for each of the pictures they have made. They write these sentences on their pictures.

This is a chance for students to discuss the relationship between addition and multiplication, and the fact that changing the order of the numbers does not change the answer—that is, 2 cars with 4 wheels each could be written either 2 × 4 or 4 × 2. Children need to be flexible in changing the order of factors to make problems easier. For example, 3 × 10 may be more familiar to them than 10 × 3. See the **Dialogue Box,** What Is Multiplication? (p. 24), for an example of encouraging flexible student thinking. The **Teacher Note,** The Relationship Between Division and Multiplication (p. 29), further discusses the importance of flexibility in writing number sentences.

During the remainder of this session, students make additional pictures that show items from the lists and number sentences that haven't been illustrated yet. You might now encourage some students to make pictures with more than 6 groups. Remind them to write both word sentences and the multiplication sentence to describe each picture.

❖ **Tip for the Linguistically Diverse Classroom** Both in class and for the homework, give students with limited English proficiency the option of writing their word descriptions in their native language; their multiplication number sentences will clarify their intentions.

Note: Be sure to save, or ask students to save, their completed pictures for use in the next session. Also save the class lists of things that come in groups for use in Investigation 4.

Session 2 Follow-Up

Homework

For homework, students finish the pictures they started in class and, perhaps, make new ones. Ask them to describe each picture both in words and with a multiplication number sentence. As necessary, send home paper for students to work on.

DIALOGUE BOX

What Is Multiplication?

This discussion happened during the activity Writing "Groups of" as Multiplication (p. 23), when the students were exploring the relationship between addition and multiplication and observed that changing the order of the numbers does not change the answer.

If I say 3 × 2, what does that really mean?

Midori: 3 × 2 would be 3 + 3 = 6. It's more like you do it twice.

Or you could think of it as 2 + 2 + 2.

Jennifer: If you have 2 apples and you get 2 more apples and then another 2 more apples, then you have 3 groups of 2, and you bring them all together and you have 6.

Jeremy: If you switch the numbers around, 3 × 2 is the same as 2 × 3.

Midori: You could think of it like groups of 3, or maybe groups of 2.

Jeremy: If the groups are of 3, then you have 2 groups.

What About Notation?

Teacher Note

It is important that your students learn to recognize, interpret, and use the standard forms and symbols for multiplication and division, both on paper and on the calculator. In this unit, students will use these:

    12   3 × 12   12 ÷ 3   3)$\overline{12}$
  × 3

Your challenge is to introduce these symbols in a way that allows students to interpret them meaningfully. That is, students must understand *what is being asked* in a problem that is written in standard notation. They can then devise their own way to find an answer. Notation is also useful as an efficient way to record a problem and its solution. It is not a directive to carry out a particular procedure, or a signal to forget everything you ever knew about the relationships of the numbers in the problem!

Your students may come to you already believing that when they see a problem like 3)$\overline{42}$, written in the familiar division format, they must carry out the traditional long-division procedure. Instead, we want them to use everything they know about these two numbers in order to solve the problem. They might skip count by 3's out loud or on the calculator. (For tips on skip counting on the calculator, see the Ten-Minute Math appendix, p. 106.) Or they might use reasoning based on their understanding of number relationships:

> It takes ten 3's to make 30. Then there are three more 3's to get up to 39, that's thirteen 3's so far. Then 40, 41, 42—that's one more 3—it's 14!

> Well, half of 42 is 21, and I can divide 21 into seven groups of 3, so you double that, and it's 14.

Similarly, when students see a multiplication problem like 4 × 55 written vertically, they are likely to forget everything they know about these numbers and try to carry out multiplication with carrying. Instead, we want students to use what they know about landmarks in the number system and other familiar number relationships. For example:

> I know that two 50's make 100, and there are four 50's, so that's 200. Then I know that four 5's is 20, so it's 220.

Students need to get used to interpreting multiplication in both horizontal and vertical form as simply indicating a multiplication situation, not a particular way to carry out the problem. So, while you help students to read standard notation and to use it to record their work, keep the emphasis on understanding the problem context and using good number sense to solve the problem. For more about division notation, see the **Teacher Note**, Talking and Writing About Division (p. 81).

Session 3

Writing and Solving Riddles

Materials

- Student pictures (from Session 2)
- Cubes or blocks (50 per student)

What Happens

Students pose problems about their pictures by leaving out one of the two factors or the answer, then solve each other's riddles. Student work focuses on:

- writing questions from mathematical statements

Activity

Writing Riddles for Our Pictures

Invite a few volunteers to show any new counting pictures they may have made for homework. Then hold three of the students' pictures in your hand without showing them to the class. Ask a riddle about each picture. Each riddle should supply two of the three pieces of descriptive information about the picture (both factors, or one factor and the total).

In this picture I see 12 wheels on cars. Each car has 4 wheels. How many cars do I see?

In this picture I see 4 sports teams. There are 36 players altogether. How many players are on each team?

In this picture there are 4 flowers. Each flower has 5 petals. How many petals do I see?

Students write their own riddles on the back of two or three of their counting pictures. Explain that each riddle should give only two of the three pieces of information about the picture, and that the riddle should end with a question about the missing piece of information. Challenge students to try to write at least two riddles for each picture.

❖ **Tip for the Linguistically Diverse Classroom** Students with limited English proficiency can invent symbolic ways to write their riddles with drawings and numbers. They will also be better able to understand their classmates' riddles if rebus drawings are included with the sentences on the back of the picture.

For students who have difficulty with this activity, you might break the process down into steps, as follows:

26 ■ *Investigation 1: Things That Come in Groups*

1. The students identify the three pieces of information in the word sentences they wrote about a picture. For example:

   There are 4 flowers.
   Each flower has 5 petals.
   There are 20 petals altogether.

2. They write each piece of information (or the whole sentence) on a separate slip of paper.

3. They lay out the three slips of paper, in order, and choose one to turn over so the writing is hidden.

4. Students can start their riddle by writing down the information from the two slips of paper that are still facing up. For example, suppose they have turned over the slip "5 petals," leaving "4 flowers" and "20 petals altogether." They could start the riddle by writing:

   There are 4 flowers. They have 20 petals altogether.

5. Finally, they look at the slip that is turned over (in the example, "5 petals"). To complete the riddle, they ask a question that would have this information for the answer:

   How many petals does each flower have?

Students can write their riddles on small pieces of paper and tape them to the back of their pictures as shown below.

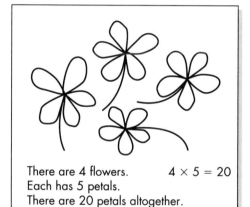

Front of student's picture.

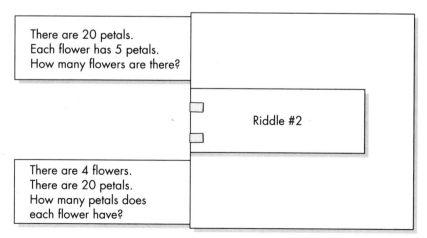

Back of picture, with riddles.

Students trade their pictures with one another (keeping them turned face down), choose a riddle, solve it, and check the answer by looking at the picture.

Since some of these riddles involve division, students are likely (at this point in the year) to solve them the same way they solve multiplication problems—as repeated addition or skip counting. See the **Teacher Note**, The Relationship Between Division and Multiplication (p. 29).

Activity

Teacher Checkpoint

Do They Understand Multiplication?

Take some time at the end of this session or the beginning of the next to observe students individually to see whether they understand the structure of multiplication problems by giving them a problem to illustrate with interlocking cubes at their desks.

Write on separate pieces of scrap paper multiplication problems that use small numbers. For example:

$2 \times 3 \quad 2 \times 4 \quad 2 \times 5 \quad 3 \times 3 \quad 3 \times 4 \quad 3 \times 5 \quad 4 \times 4$

You can repeat problems, but distribute the duplicates to students not seated near each other.

After giving each student a problem to work on, walk around the room and observe whether students can demonstrate the problems by forming groups of cubes in the correct quantities. For example, a child illustrating 3×4 could make 3 groups of 4 cubes each, or 4 groups of 3. Either answer is acceptable. After showing you the groups, the student pushes them together and counts the total (12 cubes).

Session 3 Follow-Up

 **Extension**

Riddles with Larger Numbers Students make pictures with descriptions and riddles using numbers of groups from 6 to 12. (This activity could be done as homework.) You might keep the finished pictures available in one place so that others can try solving these challenging riddles when they have free time.

The Relationship Between Division and Multiplication

Teacher Note

Multiplication and division are related operations. Both involve two factors and the multiple created by multiplying those two factors. For example, here is a set of linked multiplication and division relationships:

$8 \times 3 = 24$ $3 \times 8 = 24$
$24 \div 8 = 3$ $24 \div 3 = 8$

Mathematics educators call all of these "multiplicative" situations because they all involve the relationship of factors and multiples. Many problem situations that your students will encounter can be described by either multiplication or division. For example:

> I bought a package of 24 treats for my dog. If I give her 3 treats every day, how many days will this package last?

The elements in this problem are: 24 treats, 3 treats per day, and a number of days to be determined. This problem could be written in standard notation as either division or multiplication:

$24 \div 3 =$ ___   or $3 \times$ ___ $= 24$

Once the problem is solved, the relationships can still be expressed either as division or multiplication:

> 24 treats divided into 3 treats per day results in 8 days ($24 \div 3 = 8$)
>
> 3 treats per day for 8 days is equivalent to 24 treats ($3 \times 8 = 24$)

Many students in the elementary grades are more comfortable with multiplication than with division, just as they are often more comfortable with addition than with subtraction. We want students to recognize and interpret standard division and multiplication notation. However, we do not want to insist that they use one or the other to record their work when both provide good descriptions of a problem situation. In the dog-treat problem, either notation is a perfectly good description of the results.

Similarly, the order of the factors doesn't matter when describing a multiplication situation. Both of the examples that follow provide good descriptions of the dog-treat problem:

> 3 treats per day for 8 days is equivalent to 24 treats ($3 \times 8 = 24$)
>
> (3 per group in 8 groups is 24 total)
>
> 8 days with 3 treats per day is equivalent to 24 treats ($8 \times 3 = 24$)
>
> (8 groups with 3 per group is 24 total)

While some people prefer one or the other way to write these factors, we do not feel that a standard order (either putting the number of groups first or the number in each group first) should be taught or insisted upon. As long as students can explain their problem and their solution and can relate the notation clearly to the problem, the order of the factors in multiplication equations is not critical.

Writing and Solving Riddles ■ **29**

Session 4 (Excursion)

Each Orange Had 8 Slices

Materials

- *Each Orange Had 8 Slices*
- Calculators, cubes

What Happens

Students listen to the problems presented in the book *Each Orange Had 8 Slices* by Paul Giganti, Jr., and Donald Crews. Students track the two- and three-step problems and try to solve them on their own. Student work focuses on:

- multiplying to solve story problems
- keeping track of partial answers in two- and three-step problems.
- checking problems by skip counting

Note: If you cannot locate a copy of this book, go on to Investigation 2.

Activity

How Many Altogether?

How Many Wheels? In this activity, you are going to read aloud from the book *Each Orange Had 8 Slices*, one problem at a time. (Note that you should begin with the second problem about wheels on tricycles, because it is easier than the first and students can do it mentally.) After introducing the book, explain what students are to do.

Listen carefully. I am going to read some information and ask some questions for you to figure out. I'm not going to show you the pictures until after you have thought about the problem. As I read, *think* about the answers to the questions, but do not say them aloud.

Read the the second problem from the book, about the 3 children, their tricycles, and each tricycle's 3 wheels. Pause after each statement of fact and after each of the three questions.

After students have had time to think about the answers, ask the class to respond to each question as a group. Show the illustration in the book and ask for a volunteer to count the wheels. Then ask for someone else to count them in another way (perhaps by 3's instead of by 1's) to check.

Once the class has agreed on the answer, ask students to give a number sentence that describes the problem. ($3 \times 3 = 9$)

How Many Tiny Black Bugs? Now go back and read aloud the first problem in the book, asking about the 3 red flowers with 6 petals each, and 2 bugs on each petal. Do not show students the illustration yet. First, write the numerical information and the questions on the board:

  3 flowers
  6 petals each
  How many petals altogether?

  2 bugs on each petal
  How many bugs altogether?

Encourage students to develop whatever strategies they can to answer the questions on their own. They may use interlocking cubes or other counters, calculators, or draw or write anything that will help them. They might work with partners to try more than one way of solving the problems; they should agree on an answer.

Make sure all students understand what they are to find out and have found a way to begin. Then make note of the different ways students are working, in order to bring the variety of ways into a later discussion.

Ask students to write down their answers to the questions and caution them not to call out answers. Show the illustration from the book to the class or carry the book around to show to small groups as they finish. Allow students time to figure out the answers from the illustration by counting the petals and the bugs by 2's or 3's. They should check these answers against the ones they have written down.

Briefly bring the whole group together for students to explain how they did the two-step problem. Encourage different students to discuss different strategies. For an example of such a discussion, see the **Dialogue Box**, How Many Petals? How Many Bugs? (p. 33). Discuss the multiplication sentences that describe the two steps of the problem and the importance of labeling each partial answer in a problem with more than one step:

What multiplication sentence tells how many petals?
($3 \times 6 = 18$ petals)

What multiplication sentence tells how many bugs?
($2 \times 18 = 36$ bugs, or $3 \times 12 = 36$ bugs)

How Many Ducks and How Many "Quacks"? If time permits, challenge students with one more problem from the book—the third one, which involves groups of big waddling ducks, little baby ducks, and their "quacks."

This problem is more challenging than the first two because the relationships do not build step by step; the numbers of waddling (mother) ducks and baby ducks need to be combined to find the total number of ducks before multiplying by 3 to find the total number of "quacks." You might suggest that students do this problem with a partner or in small groups of four.

Note: In Investigation 2, there will be a Choice Time during Sessions 3 and 4 when students can work on some more problems from *Each Orange Had 8 Slices*. If you plan to offer this choice, identify some problems in the book for students to pick from.

DIALOGUE BOX

How Many Petals? How Many Bugs?

In this discussion, students are describing the different strategies they used to solve one of the two-step story problems in *Each Orange Had 8 Slices*.

How did you find the number of petals?

Annie: We just added.

How did you do that?

Annie: I wrote 6 and 6 and 6 for the petals.

Tamara: Then we added 18 and 18 to find how many bugs. I know 6 and 6 is 12 and then we added 6 more, [*counting on her fingers*] 13, 14, 15, 16, 17, 18.

Chantelle: I wrote 12 and 12 and 12 for the bugs. I added 10 and 10 and 10. That's 30. Then I added 2 and 2 and 2 is 6. That makes 36 bugs.

Michael: We used these egg cartons. Then we counted 6 holes and counted more for the second group of 6 and more for the third group of 6.

Ricardo: Then we pretended the beans are bugs. I put two beans in each egg hole.

How did you find out how many altogether?

Ricardo: I counted the beans.

Jennifer: I did what Ricardo did. I counted by 2's.

These students organized the problem in different ways. Some students used counters, while others did the computation mentally and kept track on their fingers. Michael, Ricardo, and Jennifer found the number of petals and then counted by 2's to find the number of bugs. Annie and Tamara found the number of petals and then doubled to find the number of bugs. Chantelle found the number of bugs on each flower and then added three times to find the number of bugs on three flowers.

Notice that these children counted and added to find the answers to what we think are multiplication problems. They will begin to multiply only when multiplying makes sense to them.

INVESTIGATION 2

Skip Counting and 100 Charts

What Happens

Session 1: Highlighting Multiples in 100 Charts Students highlight multiples of 2's and 3's by making a chart for each one. They discuss the patterns they find and count around the class by 2's and then 3's.

Session 2: Using the Calculator to Skip Count Students learn to skip count on the calculator and they continue highlighting multiples on the 100 charts.

Sessions 3 and 4: More Practice with Multiples Students finish highlighting their books of 100 charts and continue looking at patterns in them. In Choice Time, they work on activities that help them become more familiar with multiples.

Sessions 5 and 6: Discussing Number Patterns Students discuss number patterns in several of the charts and look at some patterns across the charts. They then use the information from all the charts to play a game, Cover 50.

Mathematical Emphasis

- Recognizing that skip counting represents multiples of the same number and has a connection to multiplication

- Finding patterns in multiples of 2, 3, 4, 5, 6, 9, 10, 11, and 12 on the 100 chart

- Understanding that the patterns numbers make can help us multiply those numbers

INVESTIGATION 2

What to Plan Ahead of Time

Materials

- Overhead projector (Sessions 1–2 and 5–6)
- Transparency pens in different colors (Sessions 1–2 and 5–6)
- Colored pencils, markers, or crayons (Sessions 1–2)
- Calculators: 1 per student (Sessions 2–4)
- Interlocking cubes: 50 per student (Sessions 1, 3–4)
- *Each Orange Had 8 Slices* (Sessions 3–4, optional)
- Legal-size envelopes, for holding Cover 50 game pieces: 1 per student (Sessions 5–6)
- Scissors (Sessions 5–6)

Other Preparation

- Duplicate student sheets (located at the end of this unit) in the following quantities:

  For Session 1

  Student Sheet 1, 100 Chart with Skip Counting Circles: 11 per student, plus some extras; also 11 overhead transparencies

  For Sessions 3 and 4

  Student Sheet 2, Patterns Across the Charts: 1 per student

  For Sessions 5 and 6

  Cover 50 Game (p. 122): 2 per student, plus some extras

  How to Play Cover 50 (p. 123): 1 per student (homework)

- Make yourself a book of highlighted 100 charts for multiples of 2 through 12, like those the students will be making (Sessions 1–6).

Session 1

Highlighting Multiples in 100 Charts

Materials

- Student Sheet 1 (11 per student)
- Interlocking cubes (available for those who want them)
- Colored pencils, markers, or crayons
- Overhead projector, transparency pens

What Happens

Students highlight multiples of 2's and 3's by making a chart for each one. They discuss the patterns they find and count around the class by 2's and then 3's. Student work focuses on:

- skip counting by 2's and 3's
- finding patterns in 2's and 3's charts

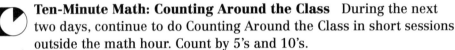**Ten-Minute Math: Counting Around the Class** During the next two days, continue to do Counting Around the Class in short sessions outside the math hour. Count by 5's and 10's.

Before you begin, ask:

Do you think our final number will be more than 50? Why do you think so? Do you think it will be more than 100? More than 200?

Stop two or three times during the count and ask questions like this:

We're at 35 now—how many more students will have to have turns to get to 50?

Activity

Highlighting 2's and 3's

Multiples of 2 Using the overhead projector, show a 100 chart and label it "2's."

I want to highlight the numbers that can make groups of 2 today. I'm going to label the top of my 100 chart "2's." What numbers do we know we can make into groups of 2 with none left over?

Highlight numbers on the 100 chart according to your students' instructions. The highlighting can be done in any way that shows the patterns clearly but leaves the numbers easy to read: circling, shading in, or outlining the frames around the numbers.

After enough numbers have been highlighted to begin to form a pattern, ask students to name multiples of 2 among the numbers over 50 (skipping around, not in order). After highlighting about half the multiples of 2, ask students what patterns they see that can help them fill out the rest of the chart. They should notice the columns of highlighted numbers under 2, 4, 6, 8, 10.

36 ■ *Investigation 2: Skip Counting and 100 Charts*

Pass out a copy of Student Sheet 1, 100 Chart with Skip Counting, to each student. Students label the top of the chart "2's" and then color in the even numbers—the multiples of 2. Suggest that students begin by marking lightly with pencil so that they can erase any mistakes. After completing about three rows, they should check with a partner to make sure they are on the right track before completing the chart with a permanent color.

Students who finish early can write 2, 4, 6, 8, and so on in the circles at the bottom of the student sheet. If there is even more time, students can turn the sheet over and write some of the patterns they see. For example:

> They're all in lines going up and down.
> They're stripes.
> They're all even.
> They all end in 0, 2, 4, 6, 8!

Multiples of 3 As you project a clean 100 chart in front of the class and label it "3's," say:

Now I'd like to make a chart about 3's. This chart will be about numbers that make groups of 3, so I am labeling it "3's."

Using a color that contrasts with the color you used for the 2's chart, highlight the multiples of 3, following your students' instructions. Again, encourage students to skip around on the chart and to name some multiples of 3 larger than 50. When you have highlighted enough multiples of 3 to see the diagonal pattern, ask:

What patterns do you see? Can you use these to help pick other numbers to highlight?

1	2	③	4	5	⑥	7	8	⑨	10
11	⑫	13	14	⑮	16	17	⑱	19	20
㉑	22	23	㉔	25	26	㉗	28	29	㉚
31	32	㉝	34	35	㊱	37	38	㊴	40
41	㊷	43	44	45	46	47	48	49	50
51	52	53	54	55	56	57	58	59	㊿
61	62	63	64	65	㊿	67	68	㊿	70
71	72	73	74	㊻	76	77	78	79	80
81	82	83	84	85	86	87	88	89	⑨⓪
91	92	㊽	94	95	96	97	98	㊾	100

Give each student another copy of Student Sheet 1. Students label this chart "3's" and again, color in or highlight in some manner the multiples of 3.

Multiples of 3 are particularly difficult for students to identify and highlight accurately. As before, caution them to begin working with pencil and check their work before using crayons or markers. See the **Teacher Note**, Students' Problems with Skip Counting (p. 39), for more information.

When they have finished highlighting the 3's, students count in unison by 3's, referring to their charts if they like. Count a second time, this time without the charts, counting quietly between the 3's:

1, 2, **3**, 4, 5, **6**, 7, 8, **9**, 10, 11, **12**, 13, 14, **15**, ...

If they have not already done so, students fill in the circles at the bottom of their sheets as they count out loud. Challenge students to practice counting by 3's and to prepare to recite the 3's in small groups, without their charts, during another math hour.

Activity

Making Books of 100 Charts

Hand out 9 additional copies of Student Sheet 1 to each student. Students staple the sheets together with those from the previous activity so that the 2's and 3's charts are the top two pages. Students label the rest of the sheets to make charts of multiples from 4 to 12.

If there is time now, students can highlight charts for multiples of 4 and 5. Tell students the class will be discussing 4's in the next session. Remind them to fill in the skip counting circles and write the patterns they see in the charts on the back of each sheet.

❖ **Tip for the Linguistically Diverse Classroom**   Students who are not yet writing in English may describe the patterns in their native language, supplementing their descriptions with visual diagrams.

Activity

Session 1 Follow-Up

Students continue to highlight their 100 charts in the stapled booklets and to find patterns through 8's. Remind them that they must remember to bring their book of charts back for class tomorrow.

38 ■ *Investigation 2: Skip Counting and 100 Charts*

Students' Problems with Skip Counting

Teacher Note

Some students have difficulty keeping track of their skip counting on the 100 chart. Here are some confusions we have noticed in classrooms:

- Some students always start on 1, no matter which number they are skip counting by.

- The count may get off by 1 because the student pauses at a circled number, then starts counting again with that number. For example, when counting by 6's, a student counts 6, 12, 18, then begins the next count on 18. After counting six more numbers (18, 19, 20, 21, 22, 23), the student lands on 23 instead of 24.

- Students sometimes follow a "false pattern" that doesn't actually work for the number they are counting by. For example, they may circle 3, 6, 9, then color straight down the columns under the 3, 6, and 9, not realizing that the 3's pattern doesn't continue in columns the way the 2's pattern does.

- Students may miscount one interval and then continue counting correctly, so that all subsequent numbers are affected by the original mistake. For example: 3, 6, 9, 12, 15, 19, 22, 25, 28 ....

Some of these difficulties are simply miscounting mistakes that anyone can make. Help students to use the pattern on their counting charts to check: Does the pattern continue consistently on the chart? Also, have students double-check each other. When two or three students compare charts, they can often find and correct their own miscounting.

However, some students may truly not understand what they are doing when they "count by 2's" or "count by 3's" on their charts. Here, using cubes as a first step will help. That is, when counting by 2's, the student makes a group of 2 cubes, then marks 2 on the chart; makes another group of 2 cubes (perhaps in a different color), and marks the total, 4; then makes another group of cubes, marks the total, 6; and so forth. Students will naturally stop using the cubes as soon as they feel comfortable with skip counting.

We have found that it's not helpful for students to use cubes to mark squares directly on the counting charts. Students can't see the numbers underneath them, and they often move a cube accidentally to a neighboring square, thereby misleading themselves about the pattern on the chart.

Session 2

Using the Calculator to Skip Count

Materials

- Student's books of 100 charts (from Session 1)
- Calculators
- Colored pencils, markers, or crayons
- Overhead projector, transparency pens

What Happens

Students learn to skip count on the calculator and they continue highlighting multiples on the 100 charts. Student work focuses on:

- identifying and using patterns to highlight multiples on the 100 chart
- making connections between skip counting and multiplication

Activity

Skip Counting 4's and More

Start the class with a discussion about useful strategies students can use when highlighting a 100 chart for multiples of 4. Work at the overhead filling in a 4's chart as students give you their ideas. Use questions like these to prompt the discussion:

What are some ways to make a chart for the 4's? Are there any patterns you can follow to check which numbers to highlight? How can you make sure your work is accurate?

After students have expressed their ideas, introduce the calculator as another tool they might use.

How could you use a calculator to figure which numbers make groups of 4?

As they work on their calculators, students can compare strategies with their neighbors. Most students will probably add, although some may multiply. Give them time to experiment before asking:

What answer do you get when you add 4 + 4 + 4 on the calculator? How about multiplying 3 × 4 ?

Tell students to press 0 + 4 = = = and to notice what happens each time they press the equals (=) key.

What happens if you press the equals key three times after pressing 0 + 4? (4, 8, 12 are displayed in succession.) **What if you start with 4 + and press the equals key four times?** (The calculator shows 16 displayed.)

40 ■ *Investigation 2: Skip Counting and 100 Charts*

Let the students explore using the repeated equals key with other starting numbers:

What if you start with 3 + and press the equals key again and again? What does the calculator show?

Count slowly by 4's in unison, with students using a calculator. Then count again by 4's, having students read every other number highlighted on their 2's chart. Finally, count a third time without calculators or charts, but counting slowly and quietly in unison to figure out the next number:

... **28**, 29, 30, 31, **32**, 33, 34, 35, **36**, ...

Counting by 5's and 10's Since most students have probably not yet completed a chart of 10's, do one quickly as a class, working at the overhead projector while students fill in the 10's chart in their books. Also fill in a transparency chart for the 5's, following student suggestions.

Ask students what patterns they see in the 10's. They are likely to mention that all the 10's end in 0. Now have them look at their 5's charts and ask what patterns they see. (The 5's end in 5 and 0.)

To demonstrate the relationship between the 5's and 10's, have the class count with you by 5's, alternately speaking softly and loudly: whisper 5, loudly say 10, whisper 15, loudly say 20, and so on. The group will be saying the 10's table, while whispering the odd numbers in the 5's table.

Students can quickly highlight these multiples on their charts if they haven't already, and write about the patterns they see. If time permits, they can also try skip counting by 5's and 10's on the calculator.

Session 2 Follow-Up

Students continue to work on their 100 charts for the different multiples, finishing through 12's. Remind them to complete the skip counting at the bottom of the charts, and to write some of the patterns they see on the back of each chart. Family members can help find patterns to write about. Tell students to be sure to bring their charts back for class tomorrow.

Homework

Sessions 3 and 4

More Practice with Multiples

Materials

- Student's books of 100 charts
- Student Sheet 2 (1 per student)
- Calculators, cubes
- *Each Orange Had 8 Slices* (optional)

What Happens

Students finish highlighting their books of 100 charts and continue looking at patterns in them. In Choice Time, they work on activities that help them become more familiar with multiples. Their work focuses on:

- recognizing multiples of the same number
- writing about patterns in the 100 charts
- becoming familiar with multiples

 Ten-Minute Math: Counting Around the Class During the next two days, continue to do this activity in short sessions outside of the math hour. Count around the class by 3's and then by 6's, with students reading from their 3's charts for both. (They will read only every other number for the 6's.)

Some days, you might have students use a calculator for this activity. If necessary, remind them to use the equals (=) key to skip count, or review the procedure that was presented in the last session. You might use one calculator and pass it around the room, while each student presses the equals key once in turn and reads the resulting number aloud. As a challenge, encourage students to try to say their number before pressing the equals key.

Activity

Choice Time: Exploring Multiples and Patterns

Four Choices During this session and the next, students may choose from three or four activities that are going on simultaneously in the classroom. These activities are designed to give students more experiences with the 100 chart, to help them develop knowledge of the multiples, and to apply this knowledge by solving problems. Encourage students to try the activities at home with their families after working on them in Choice Time.

How to Set Up the Choices Students will need to have the following available for the activities:

Choice 1: Skip Counting with a Partner—their books of 100 charts
Choice 2: Making Towers—their books of 100 charts, cubes, calculators, paper and pencil

Choice 3: Patterns Across the Charts—copies of Student Sheet 2, their books of 100 charts, paper and pencil

Choice 4: Solving Story Problems (optional)—the book *Each Orange Had 8 Slices*, paper and pencil

You may want to list the choices on the board or have students keep track of their work on their own choice lists. Students could write down the name of each activity they do, the charts (the number) they are working on, and the name of their partner (where applicable).

Briefly introduce the four activities. Choice 2 should be introduced with a whole-class demonstration; see the description for specifics.

Choice 1: Skip Counting with a Partner

Partners choose a set of multiples they want to practice from their book of 100 charts. One partner looks at the highlighted chart while the other skip counts without looking at the chart. The purpose of this activity is for both students to become more familiar with skip counting and with recognizing multiples. Partners take turns holding the chart and helping each other.

It is important that students choose their own partners to make this a comfortable and supportive activity. Emphasize that this is not a competition, and that partners are working together so that they both learn.

Students can try the following variations as they gain confidence:

- Instead of starting from the number itself (3, 6, 9…), they start from any multiple of the number smaller than 100. For example, on the 3's chart, they might start with 39, and say "39, 42, 45, …"
- Start just below 100 and count on beyond 100 (for example, 99, 102, 105 …)
- Count backwards (for example, 63, 60, 57).

Choice 2: Making Towers

Students start by picking a chart from their book of 100 charts. If they pick the 3's chart, they will make towers of 3 interlocking cubes. If they pick the 7's chart, they will make towers of 7 cubes.

Students then pick a highlighted number on the chosen chart. For example, they might randomly touch a place on the chart with their eyes closed and work with the nearest highlighted number. The challenge is to find out how many towers they need to make exactly that (highlighted) number of cubes, following this procedure:

1. Predict how many towers are needed.
2. Build and count the towers of cubes.
3. Check on a calculator.
4. Write the multiplication sentence.

To introduce the activity, demonstrate this procedure once. For example:

1. Point to 18 on the 3's chart.
   I think I'll need to make 6 towers to use 18 cubes.
2. Make 6 towers of 3, and then count the cubes (counting by 3's).
   In 6 towers, I have 3, 6, 9, 12, 15, 18 cubes.
3. Check your counting on a calculator, skip counting by 3's.
4. Write the sentence $3 \times 6 = 18$, or $6 \times 3 = 18$, or 6 towers of $3 = 18$.

Some teachers have suggested that to provide vocabulary practice, students also label the *factors* and the *multiple* in the multiplication sentence (see the **Teacher Note**, Introducing Mathematical Vocabulary, p. 46).

$$6 \times 3 = 18$$
$$\text{factor} \quad \text{factor} \quad \text{multiple}$$

Students can try several highlighted numbers on the same 100 chart. For example, how many 3's (towers of 3) are in 24? 51? 63? Remind students to record their number sentences.

Note: Most of the arithmetic the students have been doing has been multiplication. This activity provides a chance to practice division. Don't insist, however, that students use a division method to make their predictions. Skip counting up to the total is a perfectly good way for students to do division.

Choice 3: Patterns Across the Charts

Each student works with a copy of Student Sheet 2, Patterns Across the Charts. Students use their highlighted 100 charts to answer the questions on this sheet about number patterns in the various charts. They record their answers on the sheet after each question. Students can work in pairs or alone. This is a chance for students to complete their charts and fill in skip counting circles if they haven't done so, and to add more patterns they find to the back of their charts.

❖ **Tip for the Linguistically Diverse Classroom** For Student Sheet 2, pair students who are proficient in English with those who are not yet reading and writing in English. Encourage the use of pointing and head gestures to aid in understanding the questions as they look at the 100 charts together.

Note: All students should spend some time working on Student Sheet 2 before the class discussion about patterns across the charts in Session 5 or 6. Therefore, those who do not choose to do this activity during class time should complete the sheet as homework.

Choice 4: Solving Story Problems

If you did the Session 4 excursion in Investigation 1 and have the book *Each Orange Had 8 Slices*, students can choose to do additional problems now. If you have identified certain problems for students to choose from, mark them with stick-on notes or bookmarks.

For each problem, students write the important information on their paper. Then they explain their solution with writing and pictures, illustrating the problem and writing a multiplication sentence that describes it, as they did with their own problems in the previous investigation.

Teacher Checkpoint

Using the Skip Counting Circles

While students are working on the Choice Time activities in Sessions 3 and 4, walk around and observe them. Check that they can use the skip counting circles at the bottom of the 100 chart to multiply and to divide. For example, if a student is using the 6's chart, ask:

How can you find the answer to 4 × 6 in these circles?
How can you use these numbers to find out how many 6's are in 42?

Students should be able to count to the fourth circle to find 4 × 6. They should also be able to divide by counting how many circles there are up to and including the one with 42.

Sessions 3 and 4 Follow-Up

Students who have not done Choice 3: Patterns Across the Charts should complete Student Sheet 2 at home. Other students may continue working on any of the Choice Time activities at home. Remind them to take home their individual choice lists to record any work.

Teacher Note: Introducing Mathematical Vocabulary

In this unit, several important mathematical words come up naturally in discussing the activities. Introduce these words by beginning to use them yourself. Explain what you mean by them, but don't insist that students use them.

Factor When using the books of 100 charts and writing multiplication number sentences, ways of referring to the numbers that "work" in these activities can be wordy or cumbersome. As you talk about multiples in different contexts, you can gradually introduce the word *factor*. As long as the introduction of a mathematical word is preceded by activities that make its definition clear, students enjoy knowing an "adult" word to refer to a concept they have learned.

Multiple You may want to hold off on this word until *factor* is well established in the students' vocabulary, but it can be naturally introduced in Investigation 2, when students are highlighting multiples on the 100 charts.

Even and Odd These words will come up in the students' descriptions of patterns on the 100 charts. Don't assume that students know exactly what they mean by these words. Some children believe that an even number has only even factors. They might say, "No, 3 isn't a factor of 24 because 3 isn't even."

Row and Column In talking about their work with the 100 charts, students often confuse the words *row* and *column*, describing a pattern as going "down the row" rather than "down the column." This is a good opportunity to talk about the difference, since using *row* for both (as students often do) makes communication more difficult. However, do not insist that students use these words in the conventional way as long as they can explain or demonstrate what they mean. Remembering the difference can be hard, and focusing on getting the words right may obscure the good mathematical thinking a student is doing. Rather, keep using the terms yourself so that students continually hear them used correctly in context.

Other terms that may come up in this context and may need some explanation are *horizontal*, *vertical*, and *diagonal*.

Investigation 2: Skip Counting and 100 Charts

Sessions 5 and 6

Discussing Number Patterns

What Happens

Students discuss number patterns in several of the charts and look at some patterns across the charts. They then use what they know about multiples to play a game, Cover 50. Student work focuses on:

- recognizing multiples of the same number
- describing patterns in the 100 charts
- becoming familiar with multiples

 Ten-Minute Math: Counting Around the Class During the next two days, find time to do this short activity outside of math class. This time count by 9's, 10's, or 11's. Students may have those 100 charts in front of them to use as needed. Before you begin counting by one of these numbers, ask:

Will we reach 100? 200?

Stop two or three times during the count and ask questions like this:

We're at 72 now—how many students have we counted?

Materials

- Students' books of 100 charts
- Overhead projector, transparency pens
- Transparencies of Student Sheet 1 (those you have filled in, plus some clean ones)
- Cover 50 Game (2 per student)
- Scissors
- Legal-size envelopes
- How to Play Cover 50 (1 per student, homework)

Activity

Patterns in Multiples of 9 and 11

Students open their books of 100 charts to the 9's chart. As you project a clean transparency of Student Sheet 1, ask students to describe how to highlight the multiples of 9.

Where are the multiples of 9 on this chart? What patterns do you see in the 9's?

As you highlight the multiples of 9, students should describe a right-to-left diagonal that runs from 9 to 81. They should also notice the beginning of a second diagonal that starts with 90 and 99.

Then ask students to describe any other patterns that they see (the ones-place digits go down—9, 18, 27, and so on—while the tens-place digits go up). You might ask if anyone can explain why this happens. Don't worry if they can't make sense of it. You might ask again after they have talked about the 11's.

Discussing Number Patterns ■ **47**

100 Chart

1	2	3	4	5	6	7	8	⑨	10
11	12	13	14	15	16	17	⑱	19	20
21	22	23	24	25	26	㉗	28	29	30
31	32	33	34	35	㊱	37	38	39	40
41	42	43	44	㊺	46	47	48	49	50
51	52	53	㊾	55	56	57	58	59	60
61	62	㊿	64	65	66	67	68	69	70
71	㊲	73	74	75	76	77	78	79	80
㊼	82	83	84	85	86	87	88	89	㊾
91	92	93	94	95	96	97	98	㉙	100

Ask students to turn to their 11's charts and tell you how to highlight the multiples of 11. You can do this on the same overhead as the 9's, but with a different color pen or method of highlighting. After volunteers have guided you in highlighting, invite observations about any patterns they see. Then ask:

Why is the tens digit always the same as the ones digit? Why do both digits get one bigger each time?

This is a good challenge for third graders. They usually enjoy the 11's pattern, but probably can't explain why it occurs. Encourage them to think about it for the next few days and tell them you will be glad to hear any explanation they come up with.

Before moving on, count around the class by 9's and then by 11's, with students looking at their charts only when they need to. Counting above 100 will be difficult, but students can help each other do this.

Activity

Numbers That Appear on Two Charts

Multiples of 2, 3, and 6 Put the 2's chart and the 3's chart on the overhead together, one on top of the other. Before projecting two charts together, ask students to make some predictions:

What numbers do you think will be highlighted on both charts? What set of multiples are they?

Have students try to write the numbers that will appear on both charts, in order, before they see them. Then show the two combined charts and ask:

Why do you think that the numbers that are multiples of 2 and of 3 are the same numbers as the 6's chart? How is the 3's chart different from the 6's chart?

See the **Dialogue Box**, Multiples of 6 (p. 53), for a sample discussion. For counting-around practice after your discussion, you might try counting by 3's and then by 6's. The 6's are the even numbers highlighted in the 3's chart.

Multiples of 3, 5, and 15 To see if the students understand how charts for different numbers intersect, ask them to make another prediction:

What chart do you think we would see if we put the chart for multiples of 5 and the chart for multiples of 3 together?

To check their predictions, project the 5's chart on top of the 3's chart, and ask students to identify the numbers that are highlighted on both charts.

Multiples of 12 Ask what two charts you should put together if you want to see the multiples of 12. If students suggest 2 and 6, try it out. (Only the 6's chart will show, because all 6's are in the 2's chart.) When students suggest it, try 3's and 4's.

100 Chart

1	2	3	4	5	6	7	8	9	10
11	12	13	14	15	16	17	18	19	20
21	22	23	24	25	26	27	28	29	30
31	32	33	34	35	36	37	38	39	40
41	42	43	44	45	46	47	48	49	50
51	52	53	54	55	56	57	58	59	60
61	62	63	64	65	66	67	68	69	70
71	72	73	74	75	76	77	78	79	80
81	82	83	84	85	86	87	88	89	90
91	92	93	94	95	96	97	98	99	100

After students have had a chance to look at the 12's pattern on the combined 3's and 4's charts, project an unmarked chart (Student Sheet 1). Ask students to tell you what numbers to highlight for the 12's chart, making one full diagonal (12, 24, 36, 48, 60) and then asking where you should continue. (72, 84, 96) When you have highlighted all the multiples of 12, invite students to talk about the patterns they see.

Ask students to look back at their charts to see which have all the multiples of 12 highlighted. (2's, 3's, 4's, 6's—the factors of 12)

Why do you think all the 12's show up in the 3's chart and in the 2's chart? What other chart should they show up on? (Note that a student might correctly answer "the 1's chart.")

This brief discussion can serve as an introduction to the following class discussion, in which students will be comparing different pages in their book of 100 charts.

Note: You may want to introduce the game Cover 50 at this time, after discussing the patterns of the individual charts for 9's, 11's, 6's, and 12's, and leave the next activity—the discussion of patterns across the charts—until Session 6.

Activity

Discussion: Patterns Across the Charts

Remind students of the work they did on Student Sheet 2 in the previous sessions or as homework, answering questions about their books of 100 charts. Encourage them to share their observations and comparisons:

Are there any numbers that are *not* highlighted on any charts? If so, what are they? How are these numbers similar to one another?

Write on the board the numbers students name that are not highlighted on any charts. They might include some of these:

13, 17, 19, 23, 29, 31, 37, 41, 43, 47, 53, 59, 61, 67, 71, 73, 79, 83, 89, 97

Explain that these are called *prime* numbers. The definition of a prime number is that it has exactly two factors, itself and 1. Point out that 2, 3, 5, 7, and 11 are prime numbers that are highlighted—but only on their *own* charts.

What numbers are highlighted on lots of the charts?

Give a few minutes for all students to find the numbers that occur on the most charts. Walk around the room to see which numbers students are investigating. Write these on the board and identify the charts they are found on.

Why do these numbers show up on many charts? How are they similar? What sets of multiples have only odd numbers?

The answer to this last question is *none*, because an even number multiplied by any number gives an even answer. Let students consider this question for awhile, and encourage them to discuss it in small groups before settling on an answer.

Note: Students' books of highlighted 100 charts should be kept for use throughout the year. Help students find a place to keep them available for reference.

Activity

Playing Cover 50

Preparation Every student receives an envelope or small plastic bag, scissors, and two copies of the Cover 50 Game. One of these sheets will be used as the gameboard. From the other sheet, students cut the first 50 number squares. Have students remove the number 1 square and underline the 6 and the 9 so they don't confuse these numbers.

These number squares and the remainder of the cut chart are kept in the envelope or plastic bag. Students can increase the challenge of the game later by cutting out ten more numbers at a time (51–60, 61–70, and so on.)

For playing at home and for the introductory all-class game, each student needs a gameboard and a complete set of number squares, 2–50. For playing in a small group, only one gameboard and a single set of numbers for the group will be needed.

Introducing the Game After students have prepared their number squares, explain that you will play the game a few times as a class to learn how it goes. During this introduction, students play individually with their own gameboards and numbers.

Note: If you have not already done so, introduce the words *factor* and *multiple* in giving these instructions; see the **Teacher's Note**, Introducing Mathematical Vocabulary (p. 46). The number called out in each turn is a *factor*. The numbers placed on the gameboard are some of its *multiples*.

The students start by randomly choosing ten numbers and placing them face up, next to their gameboard.

Begin by naming a factor—start with a small number like 3. (Don't start with 2 because too many numbers are multiples.) Players each pick out all the multiples of 3 from among the numbers in front of them, and place the number squares over the corresponding numbers on their gameboard.

A volunteer then suggests another factor. Everyone again picks out all the multiples of that number from among the numbers left, and places them on the gameboard. If needed, students can use their highlighted 100 charts to help them find multiples.

The game continues until a player places all ten of his or her number squares on the gameboard. Check that any winners played correctly by having them name each number placed on the board and tell what number it is a multiple of.

Discussing Number Patterns ■ **51**

Playing in Small Groups Form groups of two to four students. Each group uses one gameboard and one set of number squares, 2–50. To start, each player draws ten number squares from the envelope. Students then take turns naming a factor for all players to find multiples of. The goal is to place all ten number squares on the board before other players get theirs placed.

As students become accomplished at this game, they will develop strategies to help them do this. For example, they will call factors of their own numbers that are *not* factors of other players' numbers. They may even name a particular number that they are having difficulty placing; this is, of course, the only way to get a prime number placed on the board. See the **Dialogue Box**, Cover 50 (p. 53), for examples of student thinking during the game.

Sessions 5 and 6 Follow-Up

Homework

Students can play Cover 50 at home with friends or family members. Provide students with copies of How to Play Cover 50 to take home with their gameboards and sets of number squares.

DIALOGUE BOX

Multiples of 6

This discussion resulted during the activity Numbers That Appear on Two Charts (p. 48), when students were considering a combination of the 2's and 3's charts projected together on the overhead.

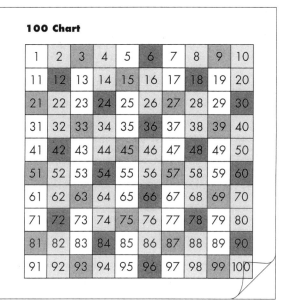

Which numbers are highlighted for both 2's and 3's?

Students: 6, 12, 18, 24 ...

Laurie Jo: It's the 6's chart!

Yoshi: The line with the 2 on top, the line with the 4 on top, the line with the 6 on top, the line with the 8 on top, and the line with the 10 on top—you go two down and highlight another one, then two down and highlight another one.

What columns are those?

Dylan: It's the even columns.

Maria: It's a number that's made out of 6's. So, it's only an even number.

Ryan: Every 30, there's a tens number.

Laurie Jo: It goes 2, 4, 6, 8, 0 [*on the diagonal*].

Sean: It goes 1, 2, 3, 4, but then it skips.

Show me what you are talking about.

Sean: In the diagonal—it's **12, 24, 36, 48,** ... then instead of 5, it's **60**.

Maria: That's the 12's [*pointing to the same 12, 24, 36 ... diagonal*].

DIALOGUE BOX

Cover 50

While playing Cover 50, these students talk about how to find all the multiples of a factor.

Elena: Multiples of 2!

Amanda: You are so sweet. [*She picks out her even numbers. There are many.*]

Aaron: OK, my turn. Multiples of 14. [*He didn't put his 14 down as a multiple of 2.*]

Elena: Aaron, you didn't have to do multiples of 14. Fourteen is a multiple of 2.

Aaron: Oh, yeah. I missed it. How do you know when you have them all?

Elena: I do the 2's by splitting the number in two piles. If it splits equally with none left over, I know 2 is a factor. And 14 makes 2 piles of 7.

Amanda: I knew because 14 ends in 4. Also I know 2 times 7 is 14.

Aaron: You know what I would do? Multiples of 1, but we would all win.

INVESTIGATION 3

Arrays and Skip Counting

What Happens

Session 1: Arranging Chairs Challenged to find different ways to arrange rows of chairs for an audience, students manipulate 12 cubes to see how many different rectangles they can make. They list the dimensions of these rectangles and the factors of 12. Students work individually and in pairs to determine the factors of other numbers by making rectangles. They also begin making sets of array cards for use throughout the investigation.

Session 2: Array Games Students talk about ways to count the total in arrays, and they learn two array games—Multiplication Pairs, and Count and Compare. In addition, students can choose to do further work on the Arranging Chairs puzzle. These activities give students practice multiplying and dividing and encourage them to develop connections between number and shape.

Session 3: The Shapes of Arrays Students briefly discuss strategies for working with arrays. Then they do an assessment problem that involves identifying by shape the arrays with a total of 36 and identifying the factors of 36. Students continue to play array games, if time permits.

Mathematical Emphasis

- Recognizing that finding the area of a rectangle is one situation where multiplication can be used
- Using arrays to skip count
- Using arrays with skip counting to multiply and divide
- Finding factor pairs
- Making connections between number and shape

INVESTIGATION 3

What to Plan Ahead of Time

Materials

- Overhead projector (Sessions 1, 2)
- Interlocking cubes: at least 15 per student (Session 1)
- Calculators (Session 1)
- Quart-size resealable plastic bags to hold array cards: 1 per student (Session 1)
- Scissors: 1 per student (Session 1)
- Manufactured Array Cards as provided in grade 3 materials kit (or use blackline masters to make your own sets)

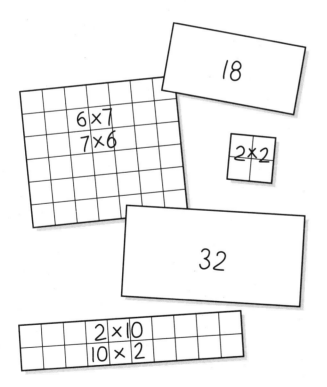

Other Preparation

- Duplicate student sheets and teaching resources (located at the end of this unit) in the following quantities:

For Session 1

Half-inch graph paper (p. 135): 2–3 per student (optional), and 1 overhead transparency

Array Cards, Sheets 1–6 (pp. 125–130): 1 set per student

How to Make Array Cards (p. 124): 1 per student (homework)

For Session 2

The Arranging Chairs Puzzle (p. 131): 1 per student (homework)

How to Play Multiplication Pairs (p. 132): 1 per student (homework)

How to Play Count and Compare (p.133): 1 per student (homework)

For Session 3

Student Sheet 3, Arrays that Total 36: 1 per student

- Make overhead transparencies of the Array Cards, Sheets 1–6. Cut apart the 51 arrays, which represent the multiplication combinations of the factors 2 through 12 with totals up to 50. Do not label the dimensions or the total.

Investigation 3: Arrays and Skip Counting ■ **55**

Session 1

Arranging Chairs

Materials

- Overhead projector
- Interlocking cubes
- Calculators
- Half-inch graph paper (optional)
- Array Cards, Sheets 1–6 (1 set per student)
- Scissors
- Resealable plastic bags
- How to Make Array Cards (1 per student, homework)
- The Arranging Chairs Puzzle (1 per student, homework)

What Happens

Challenged to find different ways to arrange rows of chairs for an audience, students manipulate 12 cubes to see how many different rectangles they can make. They list the dimensions of these rectangles and the factors of 12. Students work individually and in pairs to determine the factors of other numbers by making rectangles. They also begin making sets of array cards for use throughout the investigation. Their work focuses on:

- making rectangles for quantities of 12 and other numbers
- finding factors of 12 and other numbers

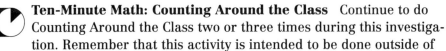
Ten-Minute Math: Counting Around the Class Continue to do Counting Around the Class two or three times during this investigation. Remember that this activity is intended to be done outside of math time.

Count by numbers whose patterns are now reasonably familiar to your students: 2's, 5's, 10's, and perhaps 3's, 4's, or 9's. Students can refer to their highlighted charts if they wish.

Ask students to predict ahead. For example, for counting by 3's, ask questions like these:

Who will say 15? Who will say 21? Khanh will be the twelfth student. What number will he say? What number will the student after Khanh say?

Ask questions about how high the counting will go.

Will we reach 50? 100? 200? What do you think will be our final number?

Arranging Chairs in Rectangular Arrays

Activity

Introducing Arrays Each student needs 12 cubes to work with. Put 12 cubes on the overhead projector. Briefly explain the task:

Here's a puzzle to solve. Pretend these 12 cubes are chairs. You want to arrange them in straight rows for an audience to watch a class play. You need to arrange the chairs so that there will be the same number in every row. How many different ways could you do this? How many chairs would be in each row? How many rows would there be? Try many different ways to arrange the chairs, even if some ways seem a bit silly for watching a class play.

❖ **Tip for the Linguistically Diverse Classroom** To support your explanation of the task, model the arranging of four chairs in different ways—one row of 4 across, four rows of 1 (one behind another), and two rows of 2. Make the corresponding arrangement of cubes for each.

Students spend some time making as many different rectangles as they can using the 12 cubes. When they have made several possible arrays, ask them to identify the number of rows and the number of chairs in each row. Show the students' different rectangles by drawing them on an overhead transparency grid, large graph paper, or on the board. Label the dimensions on each array that you show.

Identify for students the words *array* and *dimension*.

Mathematicians sometimes call things that are grouped this way to form a rectangle an *array*.

***Dimension* is a name for the length or width of a rectangle. What are the *dimensions* of your rectangles? See how I'm labeling the dimensions of the rectangles as I draw them, the *length* and the *width*.**

Use the term *by* when talking about dimensions and students will copy you; for example, "The dimensions of this rectangle are 2 *by* 6." List the pairs of dimensions on the board.

$$3 \times 4 \quad\quad 2 \times 6 \quad\quad 1 \times 12$$
$$4 \times 3 \quad\quad 6 \times 2 \quad\quad 12 \times 1$$

Have we made all of the possible rectangles? Is our list of dimensions complete? Each of the dimensions on this list is a *factor* of 12. What are all the factors of 12? (1, 2, 3, 4, 6, 12)

Activity

Arranging More Chairs

Students continue to work on the Arranging Chairs puzzle, this time with different numbers of chairs. Give each pair of students one of the following numbers to work with:

15 16 18 19 20 21 24 25 30

You might assign the numbers or conduct a drawing. If each pair does more than one number (so that all numbers are done by at least two pairs), different pairs who are working on the same number can compare their answers.

The pairs use cubes to make different arrays of chairs for their number. Then they make drawings of all the arrays they find. If you want, supply half-inch graph paper to make drawing the arrays easier. For each number they work with, students make a list of dimension pairs titled "All the Ways of Arranging ___ Chairs." They may use calculators to find or check the dimensions of their arrays.

See the **Dialogue Box**, Arranging Chairs (p. 61), for an example of a student pair trying to find all possible arrangements of 15 chairs.

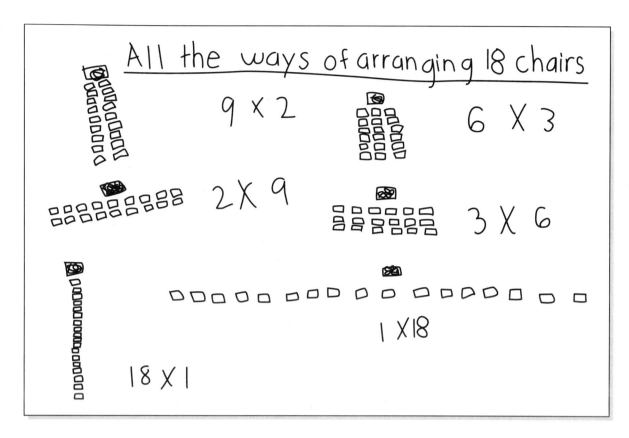

When they are finished, invite pairs of students to report their findings, one number at a time. Make a list of the dimensions of the arrays students made for each number. Point out that the number 19 makes only two arrays—1 by 19, and 19 by 1. Remind students about prime numbers—those that didn't turn up on any highlighted charts except their own. Ask:

What other numbers would have only two arrays?

Activity

Making Array Cards

The Array Cards, Sheets 1–6, provide 51 arrays—every possible array representing the multiplication equations in the 2 to 12 tables *with totals up to 50*. If you have purchased the Grade 3 materials kit for the *Investigations* curriculum, you will have printed sets of these 51 array cards that students can use in class. However, each student will also benefit from making an individual set of paper array cards, as described below, to use for homework assignments.

Give each student a set of Array Cards, Sheets 1–6, scissors, and a quart-size resealable plastic bag to hold the cut-apart array cards. Introduce the process of cutting out and labeling the cards as a whole-class activity. Give students time to practice with one or two sheets, then have them do the rest as homework or independently during the day. Emphasize that all the cards will need to be prepared before the next session.

Explain the procedure:

1. Start with Array Cards, Sheet 1. The students are to carefully cut out each individual array on the sheet, following the outlines of the grid as exactly as possible. (Seeing the exact outline of each array is important for the array games they will be playing.)

2. Students then label the grid side of each card with the dimensions of the grid.

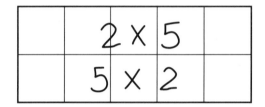

3. On the other side of each card (the blank side), students are to write the total number of squares in the grid. They may find it helpful to check the totals with a classmate or with someone at home before they write it permanently. Students may also write one of the dimensions of the grid on the total side, very lightly in pencil, to help them when the arrays are new. These can be erased when the students feel more confident.

4. Students write their initials on each card (in a corner away from the numbers) and store the labeled cards in the plastic bag.

Before students work independently, you might have them make some of the larger arrays. As students are working, walk around the room and observe whether they understand what to do and how each card should be prepared.

Session 1 Follow-Up

Students finish cutting out the array cards that they began making in class. Send home copies of the sheet How to Make Array Cards as reminders of how the cards are to be labeled. Emphasize that students are to bring their bags of cards back to class with them tomorrow.

Arranging Chairs

Ricardo and Kate are working to find all the different ways to arrange 15 "chairs" for the Arranging Chairs puzzle. They have made 3 rows of 5, 5 rows of 3, 1 row of 15, and 15 rows of 1, as their drawing shows. They are now considering whether they have all the possibilities.

Is that all there is for 15 chairs?

Kate: I think so.

How can you find out?

Ricardo: From experimenting, but nothing is even any more.

What does that have to do with it?

Ricardo: I could do like 14. Let's see, 7 + 7 is 14, so 7 + 8 … but that wouldn't be even rows, so it has to be two odds or something.

What do you know about two odds?

Kate: They make an even.

Ricardo: So, if we're missing any, it has to be odd + even. 1 + 2 doesn't work. 1 + 3 doesn't. I guess *none* of those work, because that wouldn't ever be even rows.

You see how you're starting at the beginning and making an organized list? [*Ricardo nods.*] **Well, that's what mathematicians do when they want to see if they've found all the possibilities.**

Ricardo: You're kidding!

Session 2

Array Games

Materials

- Overhead projector
- Sets of array cards
- Array card transparencies
- The Arranging Chairs Puzzle (1 per student, homework)
- How to Play Multiplication Pairs (1 per student, homework)
- How to Play Count and Compare (1 per student, homework)

What Happens

Students talk about ways to count the total in arrays, and they learn two array games—Multiplication Pairs, and Count and Compare. In addition, students can choose to do further work on the Arranging Chairs puzzle. These activities give students practice multiplying and dividing and encourage them to develop connections between number and shape. Their work focuses on:

- practicing multiplying and dividing with arrays
- connecting number and shape
- learning the multiplication tables

Activity

Counting Squares in Arrays

Tell students that during the next few days, they need to have their array cards for every math class. If they take their cards home, they must remember to bring them back the next day. (This is less important if you have purchased class sets with the *Investigations* grade 3 materials kit.)

Start this session with a discussion about ways to count the total in arrays. Choose one of your array transparencies, perhaps the 3 by 8, and put it on the overhead.

Today we are going to talk about counting the squares in arrays. What are the different ways that we could count the total number of squares in this array?

Have volunteers demonstrate at the overhead their ideas for ways to count. Some students will count by 1's, counting each individual square. Others will see that a 3 by 8 array can be counted by 3's, or perhaps by 8's. Emphasize the counting-by-groups approach, having students skip count on the arrays. Cover up all but one row of the array on the overhead, exposing one new row at a time as students skip count.

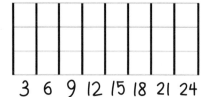
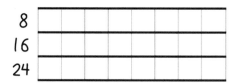

Follow this procedure with two or three more arrays that have different factors—perhaps 4 by 7, 6 by 8, and 5 by 9.

Activity

Playing Array Games

Introduce the two array games, Multiplication Pairs, and Count and Compare, to the whole class or to small groups of students. If you plan to teach the games to small groups, the rest of the class may continue finding and checking totals on their arrays or doing the Arranging Chairs puzzle while you are working with the others.

Array Game: Multiplication Pairs

The Multiplication Pairs game is designed to help students learn their multiplication tables. Students may play alone or with a partner, using one set of array cards plus notebook paper and pencil. Demonstrate the game by playing a sample round, as follows:

1. Spread out all the array cards. Some should be turned up showing the dimensions. Others should be turned over to show the total.

2. The game starts when one player chooses an array card (putting a finger on it). Emphasize that players are not to pick up the chosen card until they have given the answer. Explain that if the dimensions are showing, the player must say the total. If the total is showing, the player must give the dimensions of the grid.

3. After giving an answer, the player turns the card over to check. If the answer is correct, the player picks up the card.

4. During the game, players make lists for themselves headed "Pairs that I know" and "Pairs that I don't know yet." These lists should be kept in students' math folders.

Array Games ■ 63

The player continues to choose and identify cards (taking turns, if playing with a partner) until all the cards have been picked up.

As you are teaching students the game, point out ways they can find solutions if they get stuck.

If you get stuck on finding a total, you might think of ways to count the squares in the array. Suppose you pick an array with the dimensions 7×3 showing. Then you might count by 3's while touching each row of 3 to get the total, in this case 21.

The shape of the array can help you find the factors. Suppose you pick an array with the total 36 showing. You must figure out what the dimensions could be—6×6, or 9×4, or 12×3. In deciding what the dimensions are, the shape of the array is a good clue. If it's a square, then it must be the 6×6 array. If it's a long, thin rectangle, 12×3 is more likely.

Array Game: Count and Compare

Count and Compare is designed to give students more experience with the relationship between number and shape. The game is to be played by groups of two or three students with a single set of array cards divided among them.

Demonstrate the game by playing a sample round for two players, as follows:

1. Deal out the array cards so that both players have the same number of cards. (Set aside the one that is left over.)
2. Players place their cards in a stack in front of them *with the total side face down.*
3. Both players take the top card from their stacks and place these cards side by side (total sides still facing down).
4. Together, the players decide which array has a larger total. The player with the larger array takes both cards.

Play continues until one player runs out of cards. The arrays can then be reshuffled (with the leftover) and dealt out equally for a new game.

Sometimes arrays with the same total will be placed side by side—such as a 3×4 and a 2×6. When this happens, the players decide together who will get the cards—for example, the one whose card has the largest single dimension. Once a rule is decided, it cannot be changed until the game is over.

You may want to use the overhead projector to demonstrate how students might compare the arrays. Choose transparencies for two arrays that are close in size (such as 6×6 and 7×5), and project them side by side.

Discuss the strategies students could use to determine which array is bigger. They might compare the areas visually. They might skip count to find the total of each array. They might place one array on top of the other to compare them.

Discourage counting the squares by 1's. Be sure students understand that they should determine which array is larger without turning them over to look at the totals. When students are playing the game, challenge them to prove to each other which array is bigger.

Playing the Games   Once you have taught everyone the games, students choose between Multiplication Pairs, Count and Compare, and the Arranging Chairs puzzle (with a new number) for the rest of the class session. Explain that you expect them to play both of the two new array games so they can talk about them in the next class session.

Session 2 Follow-Up

 Homework

Students may continue to play the two array games and work on the Arranging Chairs puzzle at home. Distribute copies of the rules (The Arranging Chairs Puzzle, How to Play Multiplication Pairs, and How to Play Count and Compare, pp. 131–133) so that families can play together.

If students have only one set of array cards, remind them to bring them back to school each day.

For students who want to work on Arranging Chairs at home, you might also send home copies of half-inch graph paper that they can cut apart to record the different rectangles they make.

 **Extension**

What Number Has the Most Arrays? Challenge the class to find the number under 100 that has the most arrays. Partners work together to discuss possibilities. They may want to refer to their books of 100 charts for this, as numbers that are highlighted on many charts have the most arrays. Students can then use a calculator to investigate and narrow their choice to one or two numbers.

After a few minutes, ask for volunteers to make a list of the possible numbers on the board. Students briefly discuss the proposed numbers and choose two or three to investigate as a class. They can use calculators, pencil drawings, or interlocking cubes to figure out the different arrays. List the arrays students identify on the board. Encourage them to check the list mentally or with the calculator.

What about arrays where the rows and columns are switched—should we count them as the same or different? For example, will we count 3×2 and 2×3 as two different arrays, or as the same one?

Their decision here doesn't matter, as long as they are consistent. When you have a number of arrays listed, ask:

How can we organize these lists to tell when we have identified all the arrays? Do you think we are missing any?

Briefly discuss the results and ask students to talk about the strategies they used. Ask about what they found out:

How many arrays did each number make? Were your predictions accurate? Were there any surprises? That is, did some numbers have many more or many fewer arrays than you expected?

Some students may enjoy continuing this investigation to find other numbers that have many arrays. Encourage them to use graph paper to record all the arrays they find for a number. They may be startled to discover that large numbers do not necessarily have more factors than smaller numbers.

You might plan to post sheets for numbers that have many arrays, or other numbers that students find interesting.

66 ■ *Investigation 3: Arrays and Skip Counting*

Session 3

The Shapes of Arrays

What Happens

Students briefly discuss strategies for working with arrays. Then they do an assessment problem that involves identifying by shape the arrays with a total of 36 and identifying the factors of 36. Students continue to play array games, if time permits. Their work focuses on:

- connecting number and shape
- finding all the factors of 36

Materials

- Student Sheet 3 (1 per student)
- Sets of array cards
- Interlocking cubes

Activity

Discussing Array Game Strategies

Ask students about the strategies they developed while playing the array games. Use this discussion as another opportunity to discuss division.

When the total side of an array card is up, how do you figure out the factors or dimensions of the array?

If you have a total of 30 for an array and you know one dimension is 5, how can you figure out the other dimension?

Use this discussion as an opportunity to introduce division.

When you find one missing dimension like this, you are doing *division*. You could write this as a division problem: $30 \div 5 = x$. You also did division when you made towers of cubes a few days ago. Remember when you made towers the same size—3's or 4's or another number—to find out how many were in a larger number? Well, you were doing division then. If you were finding how many towers of 4 are in 24, how could you write that as a division problem? Decide with your partner and write it down.

Suppose we had an array card with a total of 24. What other division problems could we write, with 24 divided by one of its factors?

Collect a few ideas from students and write them on the board ($24 \div 4 = 6$, $24 \div 8 = 3$, $24 \div 2 = 12$). Then focus on the shape of the corresponding arrays.

Which of these arrays would be the most like a square? Which would be long and thin? How can you tell about the shape from looking at the numbers?

Activity

Assessment

Arrays That Total 36

Students complete Student Sheet 3, Arrays that Total 36, by labeling the dimensions of blank arrays. Students must visually consider the shape of the rectangles and the fact that the total of each array is 36. Tell them not to use rulers or any other measuring device when labeling the arrays. Suggest that they start labeling those that they know for sure first, then move on to the others.

Walk around and observe students as they work. If they seem stuck, suggest that they make a list of the factors of 36, then decide which numbers go with the square array and which go with the long, skinny one. See the **Teacher Note**, Assessment: Arrays That Total 36 (p. 69), for things to look for in student work on this problem.

Students who complete the student sheet early can play one of the array games for the rest of the class session. They may work independently or with others.

Session 3 Follow-Up

Suggest that students continue playing the array games and Arranging Chairs at home with family members or friends.

Assessment: Arrays That Total 36

Teacher Note

In the assessment problem on Student Sheet 3, there are two tasks for students: They must find the multiplication combinations that make 36 (1 × 36, 2 × 18, 3 × 12, 4 × 9, 6 × 6), and they must make a visual judgment about which dimensions work for which shapes.

The most proficient students first place the dimensions they know—usually 6 × 6, 4 × 9, and 1 × 36—and then figure out the others in comparison to these, using what they know about the relationship of shape and number. They might find 3 × 12, for example, by dividing the 6 × 6 square in half and visualizing the two halves end to end (thus doubling the 6 in one dimension, and halving it in the other dimension).

Many students figure out the dimensions of the arrays one by one, with some trial and error. They eventually find all the multiplication combinations and place them correctly, as Latisha's work below shows.

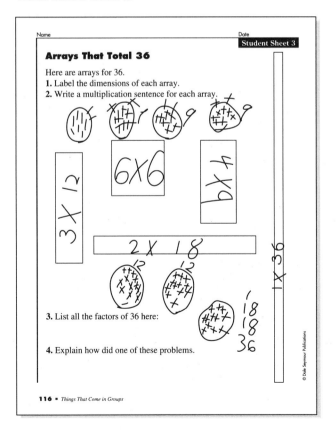

Some students will have difficulty finding factors accurately and may write incorrect equations from memory (8 × 4 = 36). You might suggest that they use their pencil to break the shape into short rows (in this case 8 short rows), label the row length (4), and skip count to see what total they get.

Some students will make little use of the different shapes as they try to figure out the dimensions. If a student has the correct factor pair for a shape, but has the factors written on the wrong sides, ask about this. For example, suppose a student has labeled one rectangle this way:

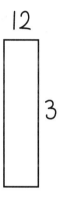

Point to the rectangle and ask which dimension is longer. In other words, if this were an array, which side which would have the most squares along the edge?

The Shapes of Arrays ■ **69**

INVESTIGATION 4

The Language of Multiplication and Division

What Happens

Sessions 1 and 2: Multiply or Divide? As students discuss the different actions suggested by division and multiplication story problems, they are introduced to division notation. They play a guessing game in which they act out multiplication and division problems. They also make class lists of multiplication and division equations for the numbers 2 through 12.

Sessions 3 and 4: Writing and Solving Story Problems Each student writes a story problem that involves multiplication or division. The class makes a book of all their problems, writing number sentences to go with each problem in the book. Students then solve each others' problems.

Mathematical Emphasis

- Understanding relationships between multiplication and division
- Identifying whether word problems can be solved using division and/or multiplication
- Using multiplication and division notation to write number sentences

What to Plan Ahead of Time

Materials

- Overhead projector (Sessions 1–2)
- Interlocking cubes in buckets (Sessions 1–4)
- Art materials: colored pencils, markers, or crayons; colored paper for book covers (Sessions 3–4)
- Calculators (Sessions 3–4)

Other Preparation

- Duplicate student sheets (located at the end of this unit) in the following quantities:

  For Sessions 1–2
  Student Sheet 4, Story Problems: 1 per student
  Student Sheet 5, What Do These Mean? 1 per student

- Make an overhead transparency of Number Problems (p. 134) (Sessions 1–2).

- Plan how to reproduce and assemble the class book of student problems (Sessions 3–4).

INVESTIGATION 4

Sessions 1 and 2

Multiply or Divide?

Materials

- Student Sheet 4 (1 per student)
- Transparency of Number Problems
- Overhead projector
- Interlocking cubes in buckets
- Class lists (from Investigation 1)
- Student Sheet 5 (1 per student)

What Happens

As students discuss the different actions suggested by division and multiplication story problems, they are introduced to division notation. They play a guessing game in which they act out multiplication and division problems. They also make class lists of multiplication and division equations for the numbers 2 through 12. Student work focuses on:

- understanding and solving multiplication and division story problems
- using multiplication and division notation to write number sentences

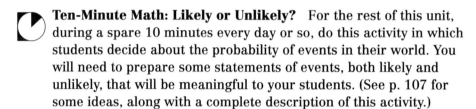

Ten-Minute Math: Likely or Unlikely? For the rest of this unit, during a spare 10 minutes every day or so, do this activity in which students decide about the probability of events in their world. You will need to prepare some statements of events, both likely and unlikely, that will be meaningful to your students. (See p. 107 for some ideas, along with a complete description of this activity.)

On chart paper, start a list with two headings: *Likely, Unlikely*. Read one statement of an event that is likely or unlikely to occur. For example:

The school bell will ring at [give the usual time] today.

Take a quick poll of students' opinions. Is this event likely or unlikely? Write the statement on the chart in the correct column. Continue with more of the statements you prepared, or ask students to suggest events for placement on the chart in either column.

Hand out a copy of Student Sheet 4, Story Problems, to each student.

Activity

Solving Story Problems

❖ **Tip for the Linguistically Diverse Classroom** Before handing out this worksheet, you might ask five students to find creative ways to make the problems understandable for students with limited English proficiency, using only the simplest English words combined with little drawings in rebus-style sentences.

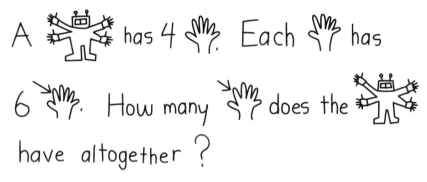

A handi-robot has 4 hands. Each hand has 6 fingers. How many fingers does the robot have altogether?

Students work individually or in pairs to solve the problem, using any method that makes sense to them. They might make drawings, or use their own hands to model the robot's hands, with pencils to represent extra fingers.

Bring students together to show how they did the problem. Invite volunteers to act out or show on the board how they solved the problem.

Students now do the remaining problems on the sheet. They may work with partners, but each student should write down enough for you to tell how they solved the problems.

Is It Multiplication or Division? When students have finished the five problems, ask them if they can tell which story problems are multiplication problems and which are division problems. Have them decide with their neighbors which problems should have multiplication sentences, and then write the number sentence for each one. If you think it's necessary, tell them that there are only two multiplication problems.

After a brief time, ask students to report on which story problems they think are multiplication problems. Volunteers can write the multiplication sentences on the board.

Note: A multiplication sentence can be used to describe all of the problems, but the "answer" to the multiplication sentence and the "answer" to the problem is the same for only problems 1 and 5. Refer to the **Teacher Note**, The Relationship Between Division and Multiplication (p. 29). See also the **Dialogue Box**, Would You Use Multiplication or Division? (p. 82), for an example of what happened in one class where there was some confusion over how to represent problem 2.

If students wrote multiplication sentences for problems 2, 3, or 4, point out that multiplication *does* describe the situation, but there is another way to write these problems more exactly—using division sentences. Remind students of the division notation.

The usual way to write a number sentence for division situations is by using a division sign. Remember when you knew the total and one dimension of an array, and you wanted to find the other dimension? If the total of the array is 20, and there are 4 squares in a row, you can write the problem as 20 divided by 4. [*Write 20 ÷ 4 = __ on the board.*] **The answer is the number of rows in the array.**

In problem 2, we have 20 muffins that must go in bags of 4, so we can write 20 divided by 4. [*Write 20 ÷ 4 = __ again.*] **The answer is the number of bags we need.**

I divide the 20 into groups of 4 because there are 4 muffins in each bag. The question I am answering in both the muffin problem and the array problem is, "How many 4's are in 20?"

Explain that the division sentence for these problems cannot be written as 4 ÷ 20. That sentence would mean 4 things divided into 20 parts, or 4 things shared among 20 people—which is possible, but not the situation in this problem.

Show students other common ways of writing the division sentence 20 ÷ 4 = __ . $4\overline{)20}$ $\frac{20}{4} =$

All these ways of writing the problem can be read as "20 divided by 4 equals what?" A good way to think of it is to ask yourself, "How many groups of 4 are in 20?"

Show students where we write the answer in each of these different forms of division notation.

Note: We do not ask students to use the form $4\overline{)20}$ in this curriculum; it is not a form used by mathematicians. However, because this form is still used on tests and in some school textbooks, students should recognize it as asking, "How many 4's are in 20?" They should be expected to find the answer in whatever way makes sense to them. For more discussion about the meaning of standard notation, read the **Teacher Notes**, Talking and

Writing About Division (p. 81) and Two Kinds of Division: Sharing and Partitioning (p. 82).

Continue with another story problem on Student Sheet 4.

Problem 3 is a little different from problem 2. In problem 2, we knew how many muffins were going into each bag. Here in problem 3, we have 12 pencils to be shared—and we don't know how many go into each group. You can think of the problem as "12 split into 3 piles. How many in each pile?" And we can write problem 3 as a division sentence. [*Write on the board 12 ÷ 3 = __.*]

Understanding a Division Problem Write two more division number sentences on the board. For example:

18 ÷ 3 = __   14 ÷ 7 = __

Students work with their partners to think of two situations that describe each problem. The first problem, for example, might involve 18 marbles that are shared among 3 friends, who will get 6 marbles each.

After a few minutes, bring the class back together and have each pair describe at least one of the situations they made up.

Activity

Acting Out Number Sentences

Project the transparency of Number Problems. This display shows groups of related multiplication and division problems (similar problems are mixed up on the page).

A. $4 \times 6 =$	B. $\begin{array}{r} 4 \\ \times 3 \\ \hline \end{array}$	C. $4\overline{)24}$	D. $\dfrac{15}{3} =$
E. $\dfrac{18}{6} =$	F. $24 \div 6 =$	G. $\begin{array}{r} 5 \\ \times 3 \\ \hline \end{array}$	H. $\dfrac{12}{4} =$
I. $12 \div 3 =$	J. $5\overline{)15}$	K. $4 \times 3 =$	L. $\begin{array}{r} 3 \\ \times 6 \\ \hline \end{array}$
M. $5 \times 3 =$	N. $3 \times 4 =$	O. $\begin{array}{r} 4 \\ \times 6 \\ \hline \end{array}$	P. $18 \div 3 =$

Demonstrate one problem with cubes on the overhead projector.

I'm going to act out, with cubes, one of the problems on this sheet. I want you to figure out which problem it is.

To do my problem, you take 18 cubes *(count out 18 cubes)*, **and you make groups of 3** *(split the cubes into 6 groups of 3).* **You make** *(point at each group as you count)* **one, two, three, four, five, six groups of 3. The answer is 6. What problem did I do?**

You have acted out problem P (18 ÷ 3 =), but it is reasonable for students to think that you acted out problem E, problem N, or problem L, because they involve the same numbers.

| E. $\frac{18}{6}$ | N. 3×6 | L. $\begin{array}{r} 3 \\ \times 6 \\ \hline \end{array}$ |

Write all student guesses on the board. Point out the similarity among problems E, L, N, and P (adding any of these to those on the board, as needed). Allow time for students to talk in pairs to plan how they would demonstrate each of these four problems a little differently.

Then act out each of the four similar problems, as students write down which is which. You may have to repeat your demonstrations several times for students to find the differences.

Following your model, student pairs pick a problem to act out, without telling the class which one it is. Allow time for pairs to choose a problem and decide together how they will act it out. They then use cubes or other small objects to show the action of multiplying or dividing that fits their problem—with minimal description, and without using the words *divide* or *multiply*. Other students guess which problem has been done.

If the class cannot agree on an answer, write down all guesses, and ask the pair to act out each of the similar problems. They should start by repeating their chosen problems, then act out the others, until the class can guess which is which.

Note: You will probably not want to have every student pair act out their problem for the class. This activity requires careful attention, and after a while, students are likely to become distracted.

Even after this activity, students may insist that some of these problems are really the same because they have the same answer. It is important to acknowledge that in multiplication, reversing the order of the factors does not change the answer. And, in fact, turning the problem around often makes it easier to solve.

Activity

Different Ways to Write Problems

Take a few minutes with the students to plan a display of the different forms of multiplication and division notations. The finished display will be posted where all the students can see it.

Start with a division problem in words, and ask students to write it with numbers in several different ways. Then ask them to write in words a multiplication problem that uses the same numbers, and also write it as number problems. Collect all ideas and write them on the display. For example:

Ways to Write Division

How many 6's in 24?

___ × 6 = 24   24 ÷ 6 =

6)$\overline{24}$ $\frac{24}{6}$ =

Ways to Write Multiplication

Six groups of 4 are 24 altogether.

$\begin{array}{r} 4 \\ \times\ 6 \\ \hline 24 \end{array}$ 6 × 4 = 24

Multiply or Divide? ■ **77**

Activity

Writing Multiplication and Division Sentences

Near the end of Session 2, return to the lists of things that come in groups that the class compiled during Investigation 1.

Now that you know how to write multiplication and division sentences, you will be adding number sentences to our lists of things that come in groups.

$1 \times 2 =$
$2 \times 2 =$
$3 \times 2 =$
$4 \times 2 =$
$5 \times 2 =$
$6 \times 2 =$
$7 \times 2 =$
$8 \times 2 =$
$9 \times 2 =$
$10 \times 2 =$
$11 \times 2 =$
$12 \times 2 =$

Demonstrate writing a list of multiplication sentences for 2's, without answers, on the board or on the overhead, or on a sheet of paper that you can attach to the posted class list.

Assign to each pair of students one of the numbers from 3 to 12, or other numbers for which you have lists of groups. Each pair is to write up a neat list of multiplication and division equations to post with the corresponding class list of things that come in groups.

For each list of equations, students start with $1 \times \_\_$ and write the sentences up to $12 \times \_\_$, as you did for the 2's. (Or, they might write the factors in the opposite order, $\_\_ \times 1$ up to $\_\_ \times 12$.) They then solve the problems, first writing the totals they know by heart, and finishing the rest using patterns in skip counting or any other method they know, with or without cubes or other counters.

After students have had time to record some of the multiplication equations, bring the class together to talk briefly about how to develop division sentences from the multiplication sentences. Point out that the answer to the multiplication sentence—or the total—becomes the first number in the division sentence—the number to be divided or shared.

As you did for multiplication, demonstrate how to write the list of division sentences for the 2's. Write the first few, without answers, on the board or on the overhead:

$2 \div 2 =$
$4 \div 2 =$
$6 \div 2 =$

Ask students to provide the next few problems by continuing the patterns they see, or by looking at the multiplication answers. Fill in these problems on the chart, writing each division sentence next to the related multiplication sentence.

Partners can then list the division facts in a similar manner for their assigned number.

Students begin to write their lists of the equations here in Session 2; they will have time in Sessions 3 and 4 to make a neat copy while you are assembling the class problem books.

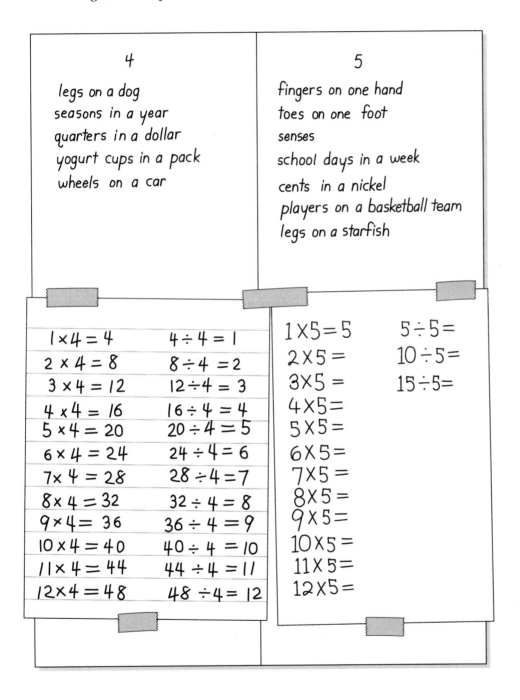

Activity

Teacher Checkpoint

Do They Understand the Notation?

To see whether students understand division and multiplication notation, take about 10 minutes of class time for students to do the problems on Student Sheet 5, What Do These Mean? Students may use cubes if they wish, but not calculators.

The numbers used in these problems are small in order to keep the focus of the work on the different forms of notation. If your students have difficulty with any of these, plan to spend additional class time studying multiplication and division notation.

Sessions 1 and 2 Follow-Up

Interpreting Problems on Standardized Tests If your students will be taking traditional standardized tests, take some time to familiarize them with pages that have a mixture of number problems without words. Display such a page on the overhead, and pick one or two problems for students to interpret in their own words and solve. Have them share their solutions.

Talking and Writing About Division

Teacher Note

Various division symbols are used as standard notation in our society:

$$24 \div 4 \qquad 4\overline{)24} \qquad 24/4 \qquad \frac{24}{4}$$

In this unit, we use only the form $24 \div 4$. However, we want students to recognize the other forms (which they may see on tests and in other textbooks) as having the same meaning as $24 \div 4$. They will use the fractional form in the Fractions units of the *Investigations* curriculum.

There are many different ways to "read" or speak of these notations:

> Four goes into (or, as students say, "guzinta") 24
>
> 24 divided by 4
>
> How many 4's are in 24?
>
> When 24 is shared among 4 people, how many does each person get?

So many symbols and so many different ways of reading them can be very confusing to young students, especially because the numbers and symbols appear in different positions, depending on which notation you are using. We would like students to read division notation with as much meaning as possible, so that they connect the symbols to the situations they represent.

The first two ways (above) of "reading" these notations correspond to the ways the symbols are written, and may seem simpler to you. However, "Four goes into 24" is a phrase that carries little meaning about the division operation. Encourage students to interpret these symbols as "24 divided into 4 parts," or "How many 4's are in 24?" These two phrases express more meaningfully what the notation represents.

Talk explicitly with your students about what these symbols mean, and how it can be confusing to read them. Work with them to help them remember that the quantity being divided is always the first number in the form $24 \div 4$, and always the inner number in the form $4\overline{)24}$. Don't let them rely on thinking of the bigger number as the one being divided, since $1 \div 2$ (for example) is a perfectly legitimate division problem, in which a smaller number is divided by a larger number.

Sometimes a problem does not divide evenly. Rather than teaching students to write "R" for the remainder, have them describe the remainder in any way that makes sense to them for that problem. For example, how many groups of 3 can be formed with 26 students? Some students may decide they can make 8 groups of 3 and one group of 2. Others may decide to make 6 groups of 3 and two groups of 4. If the example was 26 cookies to share among 3 children, they might give 8 to each and leave the remaining 2 cookies on the plate, or break them up to share.

For more about helping students connect their own good strategies with standard notation, see the **Teacher Note**, What About Notation? (p. 25).

DIALOGUE BOX

Would You Use Multiplication or Division?

A class working with the story problems on Student Sheet 4 (see discussion, p. 74) is talking about problem 2:

> We made 20 muffins for the bake sale. We put 4 muffins together in each bag to sell. How many bags of muffins did we make?

The teacher has written the following on the board:

> 20 muffins
> 4 in a bag
> How many bags?

The students are discussing possible number sentences to describe this situation.

Saloni: I can write 4 times 5 equals 20, but I don't think it's a multiplication problem.

Yes, 4 times 5 would describe this situation, but you had to already know the answer, 5 bags, to write that. How can we use just the numbers we are given, 20 and 4, to write a number sentence for the problem?

Yvonne: Is it a division problem?

Let's see. Can you think why it's a division problem? What if I multiplied 20 times 4? What would I get?

Jamal: 80.

Does that make sense? Can I have 80 bags of muffins?

Jamal: There aren't even 80 muffins, so you can't.

Yvonne: You can do 20 divided by 4.

How would we write this?

Saloni: 20, divide sign, 4.

Teacher Note — *Two Kinds of Division: Sharing and Partitioning*

There are two distinct kinds of division situations. Consider these two problems:

> I have 18 balloons for my party. After the party is over, I'm going to divide them evenly between my sister and me. How many balloons will each of us get?

> I have 18 balloons for my party. I'm going to tie them together in bunches of 2 to give to my friends. How many bunches can I make?

Each of these problems is a division situation—a quantity is broken up into equal groups. The problem and the solution for each situation can be written in standard notation as $18 \div 2 = 9$.

Although similar, these two situations are actually quite different. In the first situation, you know the number of groups—2. Your question is, "How many balloons will be in each group?" In the second situation, you know that you want 2 balloons in each group, and your question is, "How many groups will there be?" In each case you divide the balloons into equal groups, but the results of your actions look different.

Continued on next page

Teacher Note

I have 18 balloons and 2 people.
How many for each person?

I have 18 balloons to put in bunches of 2.
How many bunches?

The first situation is probably the one with which your students are most comfortable, because it can be solved by dealing out. That is, the action to solve the problem might be: one for me, one for you, one for me, one for you, until all the balloons are given out. In this situation, division is used to describe *sharing*.

In the second situation, the action to solve the problem is making groups—that is, taking out a group of 2, then another group of 2, and another, and so on, until no balloons are left. In this situation, division is used to describe *partitioning*.

Students need to recognize both of these actions as division situations and to develop an understanding that both can be written in the same way: 18 ÷ 2 = 9. Therefore, in this unit we present both kinds of division problems and, depending on the situation, help students interpret the notation as either "How many 2's are in 18?" or "Divide 18 into 2 groups; how many in each group?"

As students become more flexible with division, they will understand that they can think of a sharing situation as partitioning, or a partitioning situation as sharing, in order to make it easier to solve. For example:

How many will be in each team if I make 25 teams from 100 people?

To solve this, I will probably not deal out 100 into 25 groups. It is easier to think, "How many groups of 25 are in 100?"—knowing that the answer to this partitioning question will also give me the answer to the sharing problem I need to solve. Some of your students may soon have an intuitive understanding that they can divide in either way to solve a division problem.

Also, as students get more experience with both multiplication and division, they will begin to recognize how related problems can help them. Just as 2 × 9 = 9 × 2, 18 ÷ 2 = 9 is already related to 18 ÷ 9 = 2. When you find students stuck in routines that are difficult to keep track of—such as dealing out 100 into 25 teams—ask if thinking about a similar problem might help them. "If there were 100 people in all and 25 people on a team, how many teams would there be? Does that give you any ideas about making 25 teams from 100 people?"

Sessions 3 and 4

Writing and Solving Story Problems

Materials

- Interlocking cubes or blocks
- Calculators
- Colored pencils, markers, or crayons
- Colored paper for book covers
- Class lists from previous sessions

What Happens

Each student writes a story problem that involves multiplication or division. The class makes a book of all their problems, writing number sentences to go with each problem in the book. Students then solve each others' problems. Their work focuses on:

- solving word problems using manipulatives or drawings
- becoming familiar with relationships between multiplication and division
- writing and solving multiplication and division problems
- organizing and publishing a class book

Activity

Writing Story Problems

Explain that each student will develop and write a story problem that involves either multiplication or division. All the problems will later be published in a class book.

Students may use cubes, calculators, or drawings as needed to work out their problems. Encourage them to use complete sentences and descriptive words to make their writing interesting.

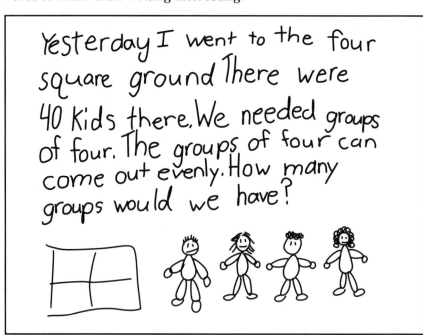

84 ■ *Investigation 4: The Language of Multiplication and Division*

❖ **Tip for the Linguistically Diverse Classroom** Give students the option of writing story problems in their native languages, or have them draw their problems, with you or another student later writing the corresponding words for them. Emphasize to the whole class that illustrations will be important to help everyone understand their problems.

As students are working, walk around the class and offer assistance as needed. Some students may need help coming up with situations that use multiplication or division rather than addition or subtraction. It is fine if they combine them, as in this particularly elaborate problem written by a third grader:

> I was going shopping at the mall. I had $39.90. I was going to buy 4 pairs of shoes and 9 packs of basketball cards. Basketball cards cost $1.75 a pack. Shoes cost eight dollars a pair. How much did the shoes cost? _____ How much did the cards cost? _____ How much money did I have left? _____

Collect your students' work and look over it to be sure the problems are clear and involve multiplication or division, before they begin work on their final drafts for publication.

Activity

A Class Book of Problems

After students have had time to develop their story problems, bring them together to talk about the class book. Explain that everyone will get a copy.

I will make copies of your story problems so we can put them together to make books for everyone. You will need to write your problem neatly and draw pictures to illustrate it. Remember to leave some room on the page for other people to solve your problem.

What shall we put on the cover of our book? What would be a good title? Should we make one cover for all the books, or would each of you like to make your own covers?

Although students should develop their problem situations during math hour, you might use writing or art time for this next step, when students write, edit, and illustrate their problems on standard white paper. If you want smaller books, provide half-size pages for students' final drafts. Let students decide how they want to do the covers; you might reproduce a page listing the title and authors, and let students illustrate and decorate their own covers.

You will need to arrange time for photocopying the pages and assembling the books. You could involve students in copying, collating, and stapling them.

During this activity, other students can continue to work on the activity begun in Sessions 1 and 2—making neat lists of the multiplication and division factors to add to the class lists of things that come in groups.

Activity

Solving Problems in the Class Book

When the books are finished and everyone has a copy, students work alone or with partners to solve the problems. They can use cubes, calculators, drawings, or anything else that will help them. If several students get really stuck on a particular problem, they may go to the author for a hint, but this is to be done only as a last resort. Remind students to label their answers (not just "7," but "7 peanuts").

Ask students to show on each page how they do the problems. Even when they can multiply in their heads, they should write down the number sentence to show what numbers they used. If some students have trouble writing division sentences, they might first write a multiplication sentence with a question mark, and then try writing the division sentence.

Note: Some students may write two-step problems, like this one:

> I went to the store and bought 12 packs of sunflower seeds. There were 4 bags of seeds in each pack. If I share them with my brother, how many bags of sunflower seeds would each of us get?

In this situation, two sentences are needed to solve the problem:

$4 \times 12 = 48$ bags, and $48 \div 2 = 24$ bags each

The problem could also be solved this way:

$12 \div 2 = 6$ packs each, and $4 \times 6 = 24$ bags each

Either way is correct.

Investigation 4: The Language of Multiplication and Division

Sessions 3 and 4 Follow-Up

Students may take home their class books and continue solving the story problems their classmates have written.

 **Homework**

Problems About the Class Ask students to make up a multiplication or division problem about their class. These problems can be serious or silly, but they should involve multiplication or division. You might begin with some examples:

 **Extension**

There are ___ students in our class. How many boxes of 6 juice bars do we need for each person to have one juice bar?

There are 4 students in one group [or use the number of students in the whole class]. Everyone in the group has two apples. Every apple has three worms. How many worms do we have in the group [or in the class]?

Start keeping a collection of problems about the class on index cards—some written by you, and some by the students. From time to time, pick one problem for students to do in a ten-minute session or for homework. Some teachers do only problems that use real data from the class:

Thirteen students bought milk today. Each of these students paid 50¢. How much milk money have I collected?

We have ___ students in our class today. How many groups of 3 students would that be? What might we do if it doesn't come out even?

You might repeat the same problem every day for a while, using the current number of students.

You can modify problems for those who find them easy. For example, if some students found the juice bars problem (above) too easy, you could challenge them to figure out how many packages would be needed to serve all the third grade classes in your school.

INVESTIGATION 5

Problems with Larger Numbers

What Happens

Session 1: Calculating Savings Students figure out how much money they would save if they saved the same amount each day for a week and then for a month. For homework, they explore what they might buy with their saved amounts. Students discuss the patterns they see in the savings tables they make.

Session 2: Many, Many Legs Students solve problems about the total number of legs on combinations of different creatures. For homework, they do a "legs" survey of living creatures in their homes or neighborhoods, and begin compiling data tables.

Session 3: Data Tables and Line Plots Students complete their data tables and calculate the total number of legs for each listed creature. They learn about line plots as another way to represent their data. Using the data tables and line plots for reference, students make up multiplication and division story problems about their data. Some of these problems will involve multiplying one-digit by two-digit numbers.

Session 4: A Riddle with 22 Legs Students work out their own ways of doing a multiplication assessment problem in which they determine all the possible combinations of spiders, cats, and people that could give us a total of 22 legs.

Mathematical Emphasis

- Multiplying and dividing in real-life situations
- Using patterns to solve multiplication and division problems
- Organizing and presenting data in tables and line plots
- Sorting out complex problems that require both multiplication and addition
- Making up division and multiplication story problems from real data

INVESTIGATION 5

What to Plan Ahead of Time

Materials

- Overhead projector (Sessions 1–2)
- Calculators (Session 3)
- Drawing paper, colored pencils, markers, or crayons
- Blank transparencies for making tables and line plots

Other Preparation

- Duplicate teaching resources and student sheets (located at the end of this unit) in the following quantities:

For Session 1

Student Sheet 6, How Much Would You Save? 1 per student; also 1 overhead transparency

For Session 2

Student Sheet 7, How Many Legs? 1 per student; also 1 overhead transparency

For Session 3

Half-inch graph paper (p. 135): at least 1 per student (optional)

For Session 4

Student Sheet 8: A Riddle with 22 Legs: 1 per student

Investigation 5: Problems with Larger Numbers

Session 1

Calculating Savings

Materials

- Student Sheet 6 (1 per student)
- Transparency of Student Sheet 6
- Overhead projector

What Happens

Students figure out how much money they would save if they saved the same amount each day for a week and then for a month. For homework, they explore what they might buy with their saved amounts. Students discuss the patterns they see in the savings tables they make. Their work focuses on:

- multiplying by 7 and 30
- using patterns to solve multiplication problems

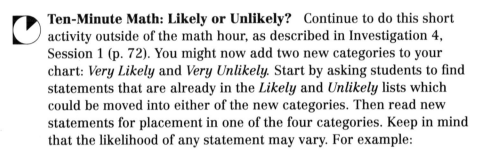

Ten-Minute Math: Likely or Unlikely? Continue to do this short activity outside of the math hour, as described in Investigation 4, Session 1 (p. 72). You might now add two new categories to your chart: *Very Likely* and *Very Unlikely.* Start by asking students to find statements that are already in the *Likely* and *Unlikely* lists which could be moved into either of the new categories. Then read new statements for placement in one of the four categories. Keep in mind that the likelihood of any statement may vary. For example:

Someone will sing during our next school variety show.

There will be a clown on the playground at recess today.

See p. 107 for a full description of this activity.

Activity

How Much Would You Save?

Ask students about their experiences with saving money.

Have any of you ever saved up your money to buy something you really wanted? What did you do? How long did it take?

Today we are going to figure out what our savings would add up to if we saved the same amount every day for an entire month.

Calculating Savings Distribute Student Sheet 6, How Much Would You Save? and show the transparency on the overhead. Ask:

If each of you saved one cent (1¢) each day for one week, how much would you have at the end of the week? How much would you have if you saved a penny every day for a month (of 30 days)?

If I save this much each day	I save this much in 1 week	I save this much in 1 month (30 days)	(HOMEWORK) What could I buy at the end of the month?
1¢	7¢	30¢	
2¢	14¢	60¢	
3¢	21¢	90¢	
4¢			
5¢			
6¢			
7¢			
8¢			

Write the amounts on the overhead transparency while students fill in the amounts in columns 2 and 3 on their sheets. Explain that students will fill out the last column as homework, investigating what the 30¢ saved in a month could buy.

Students complete the table on their own, each working on his or her own sheet. Suggest that students look for ways that patterns can help them complete the table. Encourage them to discuss with their neighbors particular strategies they use and patterns they see.

Discussing Patterns in the Savings Table After students have worked for a while, stop them briefly to share any patterns they have found.

What patterns are you using to make the work easier?

They might notice patterns like these:

- the amount-per-week column is the 7's table
- the amount-per-month column is the 3's table with a zero added to each number, because it is times 30 instead of times 3

Check students' work to see if there is any confusion about writing amounts of money. For example, students should recognize that 120¢ is written as $1.20. Some students mix the dollar and the cents signs in a single expression, and might write $1.20¢. If students have written money amounts in different ways, have them share their thinking. Then show them how to write amounts larger than one dollar with the dollar sign and decimal point.

Session 1 Follow-Up

 **Homework**

Students complete the last column on Student Sheet 6. They might go to a store, look in catalogs or advertisements, or ask family members for suggestions of what could be bought for each amount. Students list items for as many of the amounts as they can; they do not need to fill every row.

❖ **Tip for the Linguistically Diverse Classroom** Instead of writing answers in the column on the student sheet, students with limited English proficiency might bring in pictures from advertisements, cut out and pasted on another sheet of paper. They could identify each with a price and the corresponding amount saved in one month.

 **Extension**

How Many Months Old Are You? A multiplication problem that many students find interesting is changing their age in years to age in months. Explain that when people ask parents how old their babies are, they usually answer in weeks or months. They may say, "My baby is six weeks old," or "My baby is eighteen months." As the children get older—around two—the parents (and the children, too, when they can talk) begin to give their age in years.

If you gave your age in months rather than years, how old would you be? Write down your work clearly so that others can follow your reasoning.

Some students might be interested in figuring out how many *weeks* old they are, or in calculating the age in months for members of their family.

92 ■ *Investigation 5: Problems with Larger Numbers*

Session 2

Many, Many Legs

What Happens

Students solve problems about the total number of legs on combinations of different creatures. For homework, they do a "legs" survey of living creatures in their homes or neighborhoods, and begin compiling data tables. Student work focuses on:

- solving multiplication problems
- combining solutions to two or more problems

Materials

- Student Sheet 6 (completed homework)
- Student Sheet 7 (1 per student)
- Transparency of Student Sheet 7
- Overhead projector

Activity

Discussion: What Could We Buy?

Briefly discuss the Session 1 homework, asking students to share what they would buy at the end of a month for each amount calculated on Student Sheet 6. If students disagree about the cost of an item, they can tell where they got their information. This will help them realize that things cost different amounts in different stores. This discussion is likely to happen naturally outside of class, so don't spend too much class time on it.

Activity

How Many Legs?

Hand out Student Sheet 7, How Many Legs? and show the transparency on the overhead. Explain that the questions marked with a star are extra problems for anyone who finishes the others early. Read aloud the first problem, and give students a few minutes to solve it for 3 cats and 7 cats.

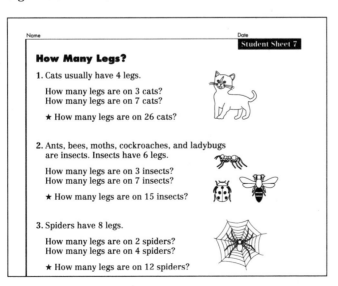

Many, Many Legs ■ 93

❖ **Tip for the Linguistically Diverse Classroom** As needed, sketch simple pictures on the overhead transparency to clarify the key words in each problem.

Check to see if everyone agrees on the answers for the cat problems. Students then work alone or with a partner to solve the remaining problems, comparing their methods and answers with partners or neighbors as needed.

When everyone has finished, students report their answers. Discuss and work out problems as a class where there is disagreement.

Throughout this investigation, you may observe students doing combinations of skip counting and multiplying. For example, Tamara determines that there are 5 four-legged animals, and she knows that $4 \times 4 = 16$, so she adds $16 + 4$ to get 20. This is a natural strategy for students who may not yet know all their multiplication tables. However, by this time in the unit, most students will know many multiples by heart. Encourage them to try to think of an answer before they start skip counting.

Activity

Planning a Survey

For homework after today's session, students will be collecting data about animals and people in their homes or neighborhoods, with numbers of legs in mind. To familiarize students with this task, ask for some ideas of what they might count. Start a table of the results. Your class discussion and resulting table might develop something like this:

> **What creatures with legs might you count in your home or in your neighborhood?**
>
> **Dominic:** Cats. There are 2 cats on my street.
>
> **Maya:** My goldfish. I have 5 goldfish with no legs each.
>
> **Latisha:** There are 4 people in my family.
>
> **Sean:** I have 2 spiders that live in my closet.
>
> **Su-Mei:** I can look out my window and see squirrels chasing each other.
>
> **How many?**
>
> **Su-Mei:** Maybe 3.
>
> **Ly Dinh:** I'd see birds—lots of pigeons—maybe 7 at once.
>
> **Let's suppose that we saw all these creatures you have mentioned in the same place. I'm going to make a table to show this information. When you are at home tonight, you will make a similar table, listing the creatures that you saw yourself.**

Write the examples that students suggested in a table on the board or overhead. Explain that students do not need to copy this table; they need only understand how to set up one like it, and how to read it.

What I saw	How many?
Cats	2
Goldfish	5
People	4
Spiders	2
Squirrels	3
Birds	7

Ask some questions about the information given by the data in the table:

How many legs would there be on all the people?

How many legs would there be on all the four-legged animals?

Students are likely to do this second question in one of two ways:

- Some will add all the creatures that have four legs, and then multiply by 4, as follows: $(2 + 3) \times 4 = 5 \times 4 = 20$
- Others will multiply the number of cats by 4, then the number of squirrels by 4, and add the results: $(2 \times 4) + (3 \times 4) = 8 + 12 = 20$

The idea that $(2 \times 4) + (3 \times 4)$ is the same thing as 5×4, or 20, is important to this discussion. If it doesn't come up naturally, bring it up.

Students who finish early might make up additional questions about this table of hypothetical data, posing problems for the whole class to solve. You or they can write the questions on the board.

Session 2 Follow-Up

Students survey their homes or a nearby place outside and make a list of the people, animals, and insects they observe for about 5 minutes during the afternoon or evening. They record these observations in a simple data table, as you demonstrated during the session. Tell them to write down the place and time that they made the observations, as well as the numbers of things they saw. They are not to record numbers of legs at this time. Remind them to be sure to bring in their work for use in the next math class.

The point of this activity is for students to collect and record some data and make up problems based on their own experiences. If any seem concerned, reassure them that it is certainly not important to make an accurate count of all the insects in or near their homes.

❖ **Tip for the Linguistically Diverse Classroom** Students with limited English proficiency might create their data tables in their native language or with drawings.

Session 3

Data Tables and Line Plots

What Happens

Students complete their data tables and calculate the total number of legs for each listed creature. They learn about line plots as another way to represent their data. Using the data tables and line plots for reference, students make up multiplication and division story problems about their data. Some of these problems will involve multiplying one-digit by two-digit numbers. Their work focuses on:

- making data tables
- making line plots
- writing story problems
- solving multiplication problems with more than one step

Materials

- Students' data tables (Session 2 homework)
- Half-inch graph paper (1 per student, optional)
- Calculators
- Drawing paper, colored pencils, markers, or crayons (optional)

Activity

Expanding Our Data Tables

To make our data tables useful for figuring out how many legs we saw running around our homes and neighborhoods, we need to add something to them.

Show students how to add two additional columns to their data table, first showing the number of legs on one creature, then the total number of legs for that type of creature. Remind students to write a sentence about when and where they collected their data if they haven't already done so.

What I saw	How many?	Number of legs on 1	Total legs
dogs	2	4	8
person	1	2	2
ants	4	6	24

These data were collected outside Rice street at 5 PM Oct. 22 by Dylan P.

To complete their expanded tables, students do the necessary computation to find the total number of legs. For students with long lists and large numbers, this will take some time. Students may help each other with this computation.

When they are finished, students exchange their tables with a partner and carefully check each other's work. They may use calculators for this.

Activity

Making a Line Plot

Gather the class for a discussion and ask a volunteer to report the data on his or her table. As the student reads, make a line plot of the data on the board. Explain what you are doing.

This is another way to show your data. It is called a line plot. It helps other people know at a glance about how many of each creature you saw.

```
                        X
                        X
     X                  X              X
     X     X     X      X       X      X
     X     X     X      X       X      X      X
    ────────────────────────────────────────────
    cats  dogs people  fish   mice  spiders  ants
```

When we look at a line plot, we can easily see what the observer saw. This line plot shows us that Yoshi counted 3 cats, 2 dogs, 2 people, 5 fish, 2 mice, 3 spiders, and 1 ant.

How many legs does one spider have? How can I use the line plot to determine how many spider legs Yoshi saw altogether?

Students might skip count by 8 as they point to each X in the spider column. Or, they could count the X's and multiply that number by 8.

As a class, practice skip counting the number of legs for a few of the creatures in the data you have displayed in the line plot. Then ask students to make up a multiplication or division problem that uses the data on the board. For example, for the line plot shown above, problems might include these:

> In this group, who has more total legs—spiders or people?
> How many legs are on all the four-legged animals?
> Are there more dog legs or cat legs?

98 ■ *Investigation 5: Problems with Larger Numbers*

What creature do we have the most of?
How many legs to they have altogether?
What kind of legs do we have the most of?

Following the model on the board, students make line plots of their own data. They can use either plain paper or graph paper.

Activity

Problems from Our Own Data

Students use their data tables and line plots for reference as they make up three or four multiplication and division problems about the data.

Encourage them to make their problems challenging. Tell them to write the answer to each problem on the back of their paper.

❖ **Tip for the Linguistically Diverse Classroom** Pair English proficient students with those who are not yet writing in English. Encourage students to use drawings and to point to their tables and line plots to aid in communication of ideas as they create their problems together.

---

1. There are 6 people in my family as shon here. If 6 of them left the house, what 3 are still there?

2. I was looking out the window of Wensday night 6:30 pm exaty. Useing my line plot can you tell how many legs in all?

[line plot with X marks over categories: people, birds, dog, cat, squirels]

1. trick question! dog, cat, squirel. The birds flew away!

2. 32 legs

Data Tables and Line Plots ■ 99

After solving the problems themselves, students exchange tables and problems with partners. Once they agree on the answers, each pair then trades problems and data with another pair of students.

If students need something clarified on someone's line plot, the creator of the plot should respond by writing additional information on the plot itself, rather than answering the other students' questions directly. This will give students valuable practice in making and reading clear data displays.

You might display students' line plots and story problems in a school hallway for students from other classes and visitors to enjoy. Students could each choose two or three problems to write with their line plot, as shown in the examples. They could also cover the answer to each problem with a flap of paper, taped in place so it can be lifted to reveal the answer.

Session 3 Follow-Up

Homework

If you are planning to make the suggested school display of line plots with related story problems, students may finish preparing their pages for this display as homework.

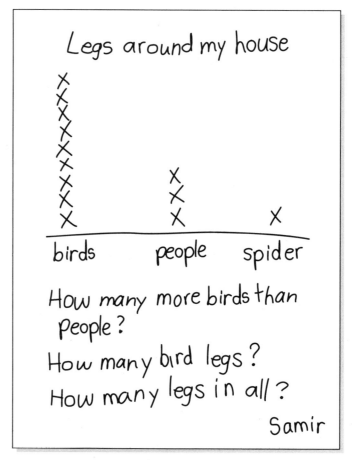

Session 4

A Riddle with 22 Legs

What Happens

Students work out their own ways of doing a multiplication assessment problem in which they determine all the possible combinations of spiders, cats, and people that could give us a total of 22 legs. Their work focuses on:

■ making a table for keeping track of possibilities
■ multiplying and dividing

Materials

■ Student Sheet 8 (1 per student, optional)

Activity

Assessment

A Riddle with 22 Legs

Pose this problem to your students:

I counted 22 legs in my house. All the legs were on cats, people, and spiders. How many of each creature—cats, people, and spiders—might be in the house? (Remember, spiders have 8 legs.)

Students will need a long time to work on this problem. Encourage them to find as many possible combinations as they can. Suggest that they list the possibilities in an organized way. They might use notebook paper to create their own recording methods, or write their answers on Student Sheet 8, A Riddle with 22 Legs.

As you review students' work, see if they have found all possible combinations (see table on p. 102).

22 Legs		
Cats	People	Spiders
5 ⑳	1 ②	0
0	11 ㉒	0
0	3 ⑥	2 ⑯
4 ⑯	3 ⑥	0
3 ⑫	1 ②	1 ⑧
	7 ⑭	1 ⑧
		1 ⑧

A Riddle with 22 Legs ■ **101**

The most proficient students will organize the data in a way that helps them keep track of the combinations they have already figured out. These students might organize all possible combinations in a chart such as the one shown below (or figure out all the combinations with 0 spiders first, then 1 spider, then 2 spiders, or organize the cats or people first).

Other students will do the computation correctly, but will not keep track of their work in a way that helps them find more solutions.

Some will struggle with the computation. Encourage those students to use a cube to represent each leg and to group them to represent different creatures.

Students may want to try other similar riddles after doing this one. Make up others using three different creatures and less than 30 legs. For example:

I counted 18 legs in my house. All the legs were on dogs, people, and ladybugs. How many of each creature—dogs, people, and ladybugs—might be in the house?

You might introduce strategies for keeping track of all possible combinations, or encourage students who seem to have developed good strategies for the assessment problem to share them with the class.

Spiders	Cats	People
2	1	1
2	0	3
1	3	1
1	2	3
1	1	5
1	0	7
0	5	1
0	4	3
0	3	5
0	2	7
0	1	9
0	0	11

Activity

Choosing Student Work To Save

As the unit ends, you may want to use one of the following options for creating a record of students' work on this unit.

- Students look back through their folders or notebooks and write about what they learned in this unit, what they remember most, what was hard or easy for them. You might have students do this work during their writing time.

- Students select one or two pieces of their best work. You also choose one or two pieces of their work to be saved in a portfolio for the year. You might include students' written solutions to the assessments in this unit—Arrays That Total 36 (p. 68) and A Riddle with 22 Legs (p. 101). Students can create a separate page with brief comments describing each piece of work.

- You may want to send a selection of work home for parents to see. Students write a cover letter, describing their work in this unit. This work should be returned if you are keeping a year-long portfolio of mathematics work for each student.

- Whatever else students save, they should keep their collections of highlighted 100 charts to be used throughout the year. These booklets will be useful for reference in the fifth unit in the *Investigations*, grade 3 sequence, *Landmarks in the Hundreds*.

Ten-Minute Math

Counting Around the Class

Basic Activity

Students count around the class by a particular number. That is, if counting by 2's, the first student says "2," the next student says "4," the next "6," and so forth. Before the count starts, students try to predict on what number the count will end. During and after the count, students discuss relationships between the chosen factor and its multiples.

Counting Around the Class is designed to give students practice with counting by many different numbers and to foster numerical reasoning about the relationships among factors and their multiples. Students focus on:

- becoming familiar with multiplication patterns
- relating factors to their multiples
- developing number sense about multiplication and division relationships

Materials

Calculators (for variation)

Procedure

Step 1. Choose a number to count by. For example, if the class has been working with quarters recently, you might want to count by 25's.

Step 2. Ask students to predict the target number. "If we count by 25's around the class, what number will we end up on?" Encourage students to talk about how they could figure this out without doing the actual counting.

Step 3. Count around the class by your chosen number. "25 ... 50 ... 75 ..." If some students seem uncertain about what comes next, you might put the numbers on the board as they count; seeing the visual patterns can help some students with the spoken one.

You might count around a second time by the same number, starting with a different person, so that students will hear the pattern more than once and have their turns at different points in the sequence.

Step 4. Pause in the middle of the count to look back. "We're up to 375 counting by 25's. How many students have we counted so far? How do you know?"

Step 5. Extend the problem. Ask questions like these:

"Which of your predictions were reasonable? Which were possible? Which were impossible?" (A student might remark, for example, "You couldn't have 510 for 25's because 25 only lands on the 25's, the 50's, the 75's, and the 100's.")

"What if we had 32 students in this class instead of 28? Then where would we end up?"

"What if we used a different number? This time we counted by 25's and ended on 700; what if we counted by 50's? What number do you think we would end on? Why do you think it will be twice as big? How did you figure that out?"

Variations

Multiplication Practice Use single-digit numbers to provide practice with multiplication facts (that is, count by 2's, 3's, 4's, 5's, 7's, and so forth). In counting by numbers other than 1, students usually first become comfortable with 2's, 5's, and 10's, which have very regular patterns. Soon they can begin to count by more difficult single-digit numbers: 3, 4, 6, and (later) 7, 8, and 9.

Landmark Numbers When students are learning about money or about our base ten system of numeration, they can count by 20's, 25's, 50's, 100's, and 1000's. Counting by multiples of 10 and 100 (30's, 40's, 600's) will support students' growing familiarity with the base 10 system of numeration.

Continued on next page

Ten-Minute Math

Making Connections When you choose harder numbers, pick those that are related in some way to numbers students are very familiar with. For example, once students are comfortable counting by 25's, have them count by 75's. Ask students how knowing the 25's will help them count by 75's. If students are fluent with 3's, try counting by 6's or by 30's. If students are fluent with 10's and 20's, start working on 15's. If they are comfortable counting by 15's, ask them to count by 150's or 1500's.

Large Numbers Introduce large numbers, such as 2000, 5000, 1500, or 10,000, so that students begin to work with combinations of these less familiar numbers.

Using the Calculator On some days you might have everyone use a calculator, or have a few students use the calculators to skip count while you are counting around the class. On most calculators, the equals (=) key provides a built-in constant function, allowing you to skip count easily. For example, if you want to skip count by 25's, you press in your starting number (let's say 0), the operation you want to use (in this case, +), and the number you want to count by (in this case, 25). Then, press the equals key each time you want to add 25. So, if you press

| 0 | + | 25 | = | = | = | = |

you will see on your screen 25, 50, 75, 100.

Special Notes

Letting Students Prepare When introducing an unfamiliar number to count with, students may need some preparation before they try to count around the class. Ask students to work in pairs to figure out, with whatever materials they want to use, what number the count will end on.

Avoiding Competition It is important to be sensitive to potential embarrassment or competition if some students have difficulty figuring out their number. One teacher allowed students to volunteer for the next number, rather than counting in a particular order. Other teachers have made the count a cooperative effort, establishing an atmosphere in which students readily helped each other, and anyone felt free to ask for help.

Related Homework Options

Counting Patterns Students write out a counting pattern up to a target number (for example, by 25's up to 500). Then they write about what patterns they see in their counting. Calculators can be used for this.

Mystery Number Problems Provide an ending number and ask students to figure out what factor they would have to count by to reach it. For example: "I'm thinking of a mystery number. I figured out that if we counted around the class by my mystery number today, we would get to 2800. What is the mystery number?"

Or, you might provide students with the final number and the factor, and ask them to figure out the number of students in the class. "When a certain class counts by 25's, the last student says 550. How many students are in the class?" Calculators can be used.

Ten-Minute Math

Likely or Unlikely?

Basic Activity

Students think about events in the world around them, considering which events are likely and which are unlikely to occur. They sort statements about events into the two categories *Likely* and *Unlikely*. As they become familiar with these ideas, adding the categories *Very Likely* and *Very Unlikely* encourages students to make finer distinctions about the probability of these events. Students can also decide whether one event is *more* likely or *less* likely than another. We avoid the categories *Certain* or *Impossible* because students of this age can get into endless arguments about whether it's indeed *certain* that the sun will rise tomorrow, or whether it's genuinely *impossible* that a large white rabbit will serve lunch in the school cafeteria today!

Likely or Unlikely? involves students in considering the likelihood of the occurrence of a particular event. Ideas about probability are notoriously difficult for children and adults. In the early and middle elementary grades, we simply want students to examine familiar events in order to judge how likely or unlikely they are. Students' work focuses on:

- describing events with terms such as *likely, unlikely, more likely*, and *less likely*
- deciding what sorts of events in their lives are more and less likely

Materials

Statements written by you or the students, naming events that are likely or unlikely. The first time you do this activity, prepare some statements ahead of time. For subsequent sessions, students can write likely/unlikely statements as part of the Ten-Minute Math session, at home, or as they arrive in class. Or, you might ask a few students to prepare some ahead. To ensure a supply of both kinds of statements, ask each student to write two, describing one likely and one unlikely event. Each statement should be written on a strip of paper that can be taped to a class list. See Step 2, below, for sample statements.

Procedure

Step 1. On chart paper, start a list with two headings, Likely and Unlikely. If this is the first time you are doing the activity, discuss with students what these words mean and what kinds of things are likely and unlikely.

Step 2. Read, one at a time, statements of events that are likely or unlikely to occur. Following are some ideas to start with. The likelihood of some of these events is, of course, related to the characteristics of your community, the season, and so forth.

> One hundred cars will pass our school during the day today.
>
> An airplane will land on our school roof today.
>
> Half of the students in our school will stay home with colds tomorrow.
>
> A few students will stay home with colds tomorrow.
>
> The school cafeteria will be noisy today.
>
> Fewer than 20 people in [our town] will order a pizza today.
>
> It will snow here tomorrow.
>
> It will rain here sometime in the next two weeks.
>
> Scientists will discover that the earth is flat.
>
> Our class will get five new students before the end of the year.

Step 3. Students decide whether the event each statement describes is likely or unlikely. After some discussion, tape the statement strip under the appropriate heading. Because there will not be enough time to discuss a statement from everyone in the class, select a few and save the rest for the next Likely or Unlikely? session. You can keep the list posted in the classroom and add new statements each time you do the activity.

Continued on next page

Ten-Minute Math

Variations

Two More Categories: Very Likely and Very Unlikely After students have had some experience with the ideas of likely and unlikely, ask them to write some statements that are *very likely* or *very unlikely*. Discuss: "How is a statement that is *very unlikely* different from one that is just *unlikely*? How is a statement that is *very likely* different from one that is just *likely*? How many of the statements we have posted already fit into these new categories? Can you think of a way to change a *likely* statement into a *very likely* statement?"

Changing Likely to Unlikely Look at your list of likely and unlikely statements. Ask students to choose one statement and change it so that it would move to the opposite list. For example:

> Unlikely: An airplane will land on our school roof tomorrow.
>
> Change to likely: An airplane will not land on our school roof tomorrow.
>
> Likely: The school cafeteria will be noisy today.
>
> Change to unlikely: The school cafeteria will be quiet today.

Choose a few of these to discuss. Do other students agree that the statement that was at first likely is now unlikely?

More or Less Likely? Introduce the element of comparison with statements using *more likely* or *less likely*. For example:

> It is more likely that it will rain tomorrow than that it will snow.
>
> It is less likely that I will see a mouse on the way home than that I will see a dog.
>
> As you or the students suggest such statements, discuss them. Do students agree with them?

Related Homework Options

Writing Likely/Unlikely Statements At home, students write statements to bring in for sorting during the next Ten-Minute Math session. You may want to provide a homework sheet with two sentences to be filled in.

> It is likely that _____ .
> It is unlikely that _____ .

After students have some experience with these ideas, you can add other sentences, making finer distinctions or comparisons:

> It is very likely that _____ .
> It is very unlikely that _____ .
> It is more likely/less likely that _____
> than that _____ .

Connections with Other Events in the Community Students might write statements of likely or unlikely events that occur in their community. For example, they might consider these statements:

> The river will flood this year.
>
> In a few years, we will have less polluted air in our city.
>
> The trash in the park will be cleaned up by next Sunday.
>
> The new school addition will be finished by September.

They may need to interview some people who know about these events to help them decide whether they are likely or unlikely. They may even be able to set into motion actions that could change the probability of some event, such as organizing a park cleanup!

VOCABULARY SUPPORT FOR SECOND-LANGUAGE LEARNERS

The following activities will help ensure that this unit is comprehensible to students who are acquiring English as a second language. The suggested approach is based on *The Natural Approach: Language Acquisition in the Classroom* by Stephen D. Krashen and Tracy D. Terrell (Alemany Press, 1983). The intent is for second-language learners to acquire new vocabulary in an active, meaningful context.

Note that *acquiring* a word is different from *learning* a word. Depending on their level of proficiency, students may be able to comprehend a word upon hearing it during an investigation, without being able to say it. Other students may be able to use the word orally, but not read or write it. The goal is to help students naturally acquire targeted vocabulary at their present level of proficiency.

We suggest using these activities just before the related investigations. The activities can also be led by English-proficient students.

Investigation 1

question, statement, illustrate

1. Show students an advertising circular that has illustrations of items for sale. Tell them that you are going to make some *statements* about the items (as you point to them):

   This is a jacket for sale.

   The jacket costs $49.

   This is a watch for sale.

   The watch costs $20.

2. Now tell students that you are going to ask some *questions* about the items (as you point to them):

   Is this jacket for sale?

   How much does it cost?

3. Tell students to listen carefully as you make statements and ask questions. If you make a statement, they are to stand up; if you ask a question, they are to sit down.

   How many kinds of shirts are for sale?

   Three different shirts are for sale.

   They come in three sizes.

   How much do they cost?

4. Explain that the drawings of the items for sale are called *illustrations*. Explain that the illustrator makes drawings that show in pictures what the words are telling about; the artist *illustrates* the words.

5. Help the students find key words in the advertising piece; challenge students to copy and then illustrate these words.

Investigation 2

chart, row, column

1. Show students a 100 chart. Point out that it shows the numbers carefully organized. Use the words *row* (horizontal) and *column* (vertical) as you point them out.

2. Make a simple chart of students' names, listing them in about three columns and four rows. Challenge students to point out the *row* that their name is in. Other students can point out the *column* where their name is listed.

   See also the **Teacher Note**, Introducing Mathematical Vocabulary (p. 46), for a discussion of using the terms *row* and *column* in the classroom.

calculator, press, equals key, plus key

1. Identify a calculator by name. Demonstrate how to *press* numbers to make them appear on the screen. Verbalize your actions:

   When I press the number 4, a 4 shows up on the screen.

2. Create action commands that require students to press numbers on the calculator.

   Press the number 9 on your calculator.

   Press 38 on your calculator.

3. Identify the plus and equals keys. Create simple math problems that require students to follow your verbal instructions.

   Press the number 4. Press the plus key.

   Press the number 7. Press the equals key.

   What is the answer?

Appendix: Vocabulary Support for Second-Language Learners

Investigation 5

day, week, month

1. Show students a yearly calendar. Flip the pages as you name and identify each *month*.
2. Turn to the current month. Count out the number of *days* in a *week*. Name each of the days in a week.
3. Challenge students to demonstrate comprehension of these words by answering questions that require only a single-word response.

   How many days are in a week? How many days are in this [current] month?

amount

1. With students watching, count out about 27¢ in real or plastic coins. Add and take away coins one at a time as you say how much each new *amount* is.

   This amount is 25¢. This amount is 15¢.

   This amount is 16¢. This amount is 11¢.
2. Challenge students to answer questions related to simple amounts:

   I have this [*point to dime*]**. What amount of money do I have?** [*Point to nickel and two pennies*]**. What amount of money is this?**

creature, leg, -legged

1. Point to each of your legs as you identify them.
2. Draw or show a picture of a cat and a dog. Point to and count their legs, and explain that they are 4-legged creatures.
3. Draw two spiders. Refer to each spider as an 8-legged creature. Ask students to decide the total number of legs for both spiders.
4. Draw on the board, or ask students to draw, a person, an insect, a dog or cat, a bird, and a fish to show different numbers of legs (2, 6, 4, 2, 0).
5. Ask students to demonstrate comprehension of the words *3-legged* (*4-legged, 2-legged,* and so on) *creature* by following action commands.

   Clap your hands when I point to a 2-legged creature.

   Stand up when I point to a 4-legged creature.

   Point to a 6-legged creature.

Blackline Masters

Family Letter
Student Sheet 1, 100 Chart with Skip Counting Circles
Student Sheet 2, Patterns Across the Charts
Student Sheet 3, Arrays That Total 36
Student Sheet 4, Story Problems
Student Sheet 5, What Do These Mean?
Student Sheet 6, How Much Would You Save?
Student Sheet 7, How Many Legs?
Student Sheet 8, A Riddle with 22 Legs
Cover 50 Game
How to Play Cover 50
How to Make Array Cards
Array Cards, Sheets 1–6
The Arranging Chairs Puzzle
How to Play Multiplication Pairs
How to Play Count and Compare
Number Problems
Half-Inch Graph Paper

_____ , 19 ___

Dear Family,

During the next few weeks, your child will be working on a mathematics unit about multiplication and division, called *Things That Come in Groups*.

Your child will be making lists of items that come grouped in different amounts—things like 2 shoes in a pair, 7 days in a week, 12 eggs in a carton. Later the class will use these lists to write their own story problems.

We will work with the 100 chart, which shows all the numbers 1 to 100. On this chart, your child will discover patterns in the multiples of a given number. We will also use "arrays"—objects arranged in rows and columns to form rectangles of different shapes. We play games with array cards, learning to recognize their dimensions and the total number of small squares in each. This is another way of becoming familiar with the multiplication tables.

At this point, your child will not be memorizing the multiplication tables. Through our activities, the class will naturally learn some multiples by heart. But they will also do "skip counting" and use that to find multiples. For example, counting by 3's—3, 6, 9, 12, and so on—gives us the multiples of 3.

The emphasis of this unit is on understanding what multiplication and division mean. The children will be asked to make sense of different multiplication and division situations. They need to develop their own ways for thinking about and writing about these. Family members can help with many of the assignments during this unit. For example, you can help your child look for things that come grouped in regular amounts. You can take turns skip counting on the 100 chart with your child. And you can play the number games that your child brings home.

Sincerely,

Name Date
Student Sheet 1

100 Chart with Skip Counting Circles

1	2	3	4	5	6	7	8	9	10
11	12	13	14	15	16	17	18	19	20
21	22	23	24	25	26	27	28	29	30
31	32	33	34	35	36	37	38	39	40
41	42	43	44	45	46	47	48	49	50
51	52	53	54	55	56	57	58	59	60
61	62	63	64	65	66	67	68	69	70
71	72	73	74	75	76	77	78	79	80
81	82	83	84	85	86	87	88	89	90
91	92	93	94	95	96	97	98	99	100

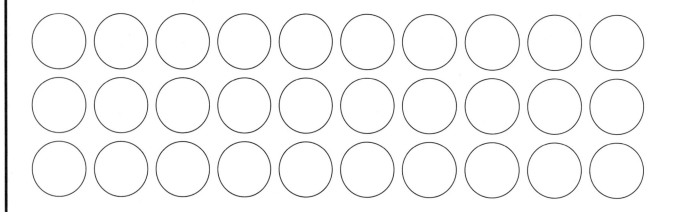

Things That Come in Groups

Name   Date

Patterns Across the Charts

Use your 100 charts to answer these questions. Discuss your answers in your group.

1. Look at the multiples of 12. What other charts have these multiples highlighted? Why do you think this is?

2. What numbers are *not* highlighted on any charts? How are these numbers similar?

3. What numbers are highlighted on *many* charts?

4. What sets of multiples have only *even* numbers?

5. What sets of multiples have only *odd* numbers?

6. What else have you noticed while answering these questions?

Things That Come in Groups ▪ 115

Arrays That Total 36

Here are five arrays for 36.
1. Label the dimensions of each array.
2. Write a multiplication sentence for each array.

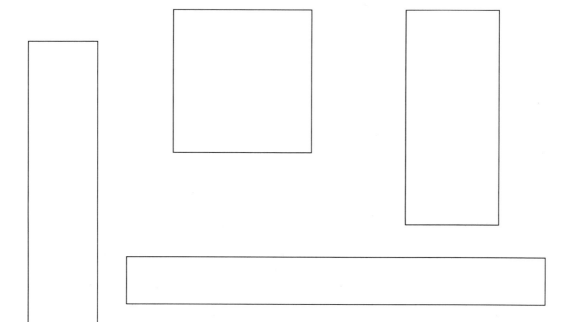

3. List all the factors of 36 here:

4. Explain how you did one of these problems.

Student Sheet 4

Story Problems

1. A handi-robot has 4 hands. Each hand has 6 fingers. How many fingers does the robot have altogether?

2. We made 20 muffins for the bake sale. We put 4 muffins together in each bag to sell. How many bags of muffins did we make?

3. Before school, my mother gave me a pack of 12 new pencils. When I get to school, I want to divide them equally between me and my two friends. How many pencils will each of us get?

4. Three children have a book of 15 movie tickets to share. Each movie costs one ticket. How many movies can each child see?

5. We bought 5 packs of pudding cups. Each pack was a different flavor. Each pack had four pudding cups. How many pudding cups did we buy?

What Do These Mean?

A. $\begin{array}{r} 4 \\ +6 \\ \hline \end{array}$	B. $12 + 4 =$
C. $4\overline{)24}$	D. $\dfrac{15}{3} =$
E. $\begin{array}{r} 15 \\ -6 \\ \hline \end{array}$	F. $3 \times 6 =$
G. $12 \div 3 =$	H. $15 - 5 =$
I. $\begin{array}{r} 5 \\ \times 3 \\ \hline \end{array}$	J. $\begin{array}{r} 18 \\ +5 \\ \hline \end{array}$

Name _____ Date _____

Student Sheet 6

How Much Would You Save?

Finish this table. Fill in the amounts you could save.
Do the last column for homework.

If I save this much each day	I save this much in 1 week	I save this much in 1 month (30 days)	(HOMEWORK) What I could buy at the end of the month
1¢			
2¢			
3¢			
4¢			
5¢			
6¢			
7¢			
8¢			
9¢			
10¢			
25¢			
50¢			
$1.00			

Things That Come in Groups ■ 119

How Many Legs?

1. Cats usually have 4 legs.

    How many legs are on 3 cats?
    How many legs are on 7 cats?

    ★ How many legs are on 26 cats?

2. Ants, bees, moths, cockroaches, and ladybugs are insects. Insects have 6 legs.

    How many legs are on 3 insects?
    How many legs are on 7 insects?

    ★ How many legs are on 15 insects?

3. Spiders have 8 legs.

    How many legs are on 2 spiders?
    How many legs are on 4 spiders?

    ★ How many legs are on 12 spiders?

4. There are 28 legs, and they all belong to cats. How many cats are there?

    ★ How many cats would there be if there were 144 legs in all?

5. We counted 3 insects, 2 cats, and 4 people in the house. How many legs are there altogether?

Name _____ Date _____

Student Sheet 8

A Riddle with 22 Legs

I counted 22 legs in my house.
All the legs were on cats, people, and spiders.

How many of each creature—cats, people, and spiders—might be in the house?

See how many different ways you can answer this riddle.
There are many possible answers.
How many can you find?

How do you know you have all the possible answers?

Spiders	Cats	People

Things That Come in Groups ■ 121

Cover 50 Game

1	2	3	4	5	6	7	8	9	10
11	12	13	14	15	16	17	18	19	20
21	22	23	24	25	26	27	28	29	30
31	32	33	34	35	36	37	38	39	40
41	42	43	44	45	46	47	48	49	50
51	52	53	54	55	56	57	58	59	60
61	62	63	64	65	66	67	68	69	70
71	72	73	74	75	76	77	78	79	80
81	82	83	84	85	86	87	88	89	90
91	92	93	94	95	96	97	98	99	100

Things That Come in Groups

How to Play Cover 50

Materials

- Cover 50 gameboard
- One set of number squares, 2–50, in a bag or envelope

Players: 2, 3, or 4

How to Play

1. Place the gameboard in the center of play. Each player draws ten number squares out of the bag or envelope.

2. Players arrange the number squares face up in front of them. Each player should be able to see everyone's number squares.

3. The player with the smallest number begins. This player calls out any factor.

4. Players search their number squares for numbers that are multiples of the named factor. Players then place these number squares over the same number on the gameboard.

5. Players take turns calling out factors and placing multiples of that factor on the gameboard. The game ends when a player has no more number squares.

How to Make Array Cards

What You Will Need

- Copies of Array Cards, Sheets 1–6
- A plastic bag to keep the cards in
- Scissors, pencil

What to Do

1. Cut out each array on the Array Card sheets. Cut them very carefully. Follow the outlines of each array as exactly as you can.

2. Each card has two sides. On the array side of the card, write the **dimensions** of the array.
   Example:

	2 × 5	
	5 × 2	

3. On the other side (the blank side), write the total number of squares in the array.
   Example:

   Also write one of the dimensions of the array, very lightly, on the card. Use pencil so you can erase later.

4. Write your initials on each card. Put them in the plastic bag. Keep your cards together at all times!

ARRAY CARDS SHEET 1

Things That Come in Groups ■ **125**

ARRAY CARDS SHEET 2

126 ■ *Things That Come in Groups*

ARRAY CARDS SHEET 3

ARRAY CARDS SHEET 4

128 ■ *Things That Come in Groups*

ARRAY CARDS SHEET 5

Things That Come in Groups ■ 129

ARRAY CARDS SHEET 6

The Arranging Chairs Puzzle

What You Will Need

30 small objects to use as chairs (for example, cubes, blocks, tiles, chips, pennies, buttons)

What to Do

1. Choose a number between 4 and 30. (Later you may use larger numbers.)
2. Figure out all the ways you can arrange that many chairs. Each row must have the same number of chairs. Your arrangements will make rectangles of different sizes.
3. Write down the dimensions of each rectangle you make.
4. Choose another number and start again. Be sure to make a new list of dimensions for each new number.

Example:
All the ways to arrange
12 chairs

Dimensions
1 by 12
12 by 1
2 by 6
6 by 2
3 by 4
4 by 3

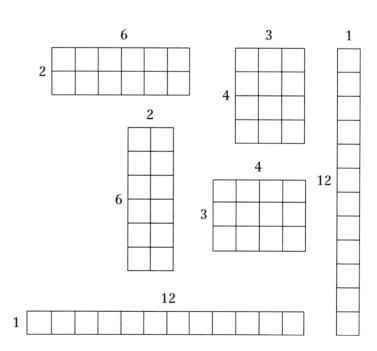

Things That Come in Groups ■ 131

How to Play Multiplication Pairs

Materials

- Set of array cards
- Paper and pencil

Players: 1 or 2

How to Play

1. Spread out all of the array cards front of you. Some should be turned up, showing the dimensions. Others should be turned over to show the total.

2. Choose an array card and put your finger on it. (Don't pick it up until you say the answer.) If the dimensions are showing, you must give the total. If the total is showing, you must say the dimensions of the grid. The shape of the array will help you.

   For example: Suppose you pick an array with the total 36 showing. The dimensions could be 6×6, or 9×4, or 12×3. You must decide which is right. The shape of the array is a good clue.

3. Turn the card over to check your answer. If your answer is correct, then pick up the card.

4. If you are playing with a partner, take turns choosing and identifying cards. Play until you have picked up all the cards

   While you are playing, make lists for yourself of "pairs that I know" and "pairs that I don't know yet." Keep these lists in your math folder.

How to Play Count and Compare

Materials: Set of array cards

Players: 2 or 3

How to Play

1. If you are playing with a partner, sit across from each other. If three people are playing, sit in a circle.

2. Deal out the array cards. Players should all have the same number of cards. Set aside any that are left over.

3. Place your cards in a stack in front of you, with the total side face down.

4. Players take the top card from their stacks and place these cards side by side (total sides still face down).

5. Decide which array is largest. You can do this by looking, by stacking them to compare, or by skip counting by rows to find the total of each. Counting the squares by 1's is not allowed.

6. The player with the largest array takes the cards, after proving that it is the largest.

7. Sometimes arrays of the same size may be played in one turn—like this:

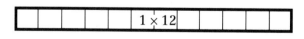

When this happens, the players decide together who will get the cards. Once a rule is decided, it cannot be changed until the game is over.

8. The game is over when one player runs out of cards.

Number Problems

A. $4 \times 6 =$	B. $\begin{array}{r} 4 \\ \times 3 \\ \hline \end{array}$	C. $4\overline{)24}$	D. $\dfrac{15}{3} =$
E. $\dfrac{18}{6} =$	F. $24 \div 6 =$	G. $\begin{array}{r} 5 \\ \times 3 \\ \hline \end{array}$	H. $\dfrac{12}{4} =$
I. $12 \div 3 =$	J. $5\overline{)15}$	K. $4 \times 3 =$	L. $\begin{array}{r} 3 \\ \times 6 \\ \hline \end{array}$
M. $5 \times 3 =$	N. $3 \times 4 =$	O. $\begin{array}{r} 4 \\ \times 6 \\ \hline \end{array}$	P. $18 \div 3 =$